Beginning Azure Synapse Analytics

Transition from Data Warehouse to Data Lakehouse

Bhadresh Shiyal

Apress®

Beginning Azure Synapse Analytics: Transition from Data Warehouse to Data Lakehouse

Bhadresh Shiyal
Mumbai, India

ISBN-13 (pbk): 978-1-4842-7060-8 ISBN-13 (electronic): 978-1-4842-7061-5
https://doi.org/10.1007/978-1-4842-7061-5

Managing Director, Apress Media LLC: Welmoed Spahr
Acquisitions Editor: Smriti Srivastava
Development Editor: Laura Berendson
Coordinating Editor: Shrikant Vishwakarma

Cover designed by eStudioCalamar

Cover image designed by Pexels

Distributed to the book trade worldwide by Springer Science+Business Media LLC, 1 New York Plaza, Suite 4600, New York, NY 10004. Phone 1-800-SPRINGER, fax (201) 348-4505, email orders-ny@springer-sbm. com, or visit www.springeronline.com. Apress Media, LLC is a California LLC and the sole member (owner) is Springer Science+Business Media Finance Inc (SSBM Finance Inc). SSBM Finance Inc is a **Delaware** corporation.

For information on translations, please e-mail booktranslations@springernature.com; for reprint, paperback, or audio rights, please e-mail bookpermissions@springernature.com, or visit http://www.apress. com/rights-permissions.

Apress titles may be purchased in bulk for academic, corporate, or promotional use. eBook versions and licenses are also available for most titles. For more information, reference our Print and eBook Bulk Sales web page at http://www.apress.com/bulk-sales.

Any source code or other supplementary material referenced by the author in this book is available to readers on GitHub via the book's product page, located at www.apress.com/978-1-4842-7060-8. For more detailed information, please visit http://www.apress.com/source-code.

Printed on acid-free paper

Table of Contents

*Dedicated to my wife, Priti,
and my daughter, Jiya.*

About the Author

Bhadresh Shiyal is an Azure data architect and Azure data engineer. For the past seven years, he has been working with a large multinational IT corporation as solutions architect. Prior to that, he spent almost a decade in private- and public-sector banks in India in various IT positions working on several Microsoft technologies. He has 18 years of IT experience, including working for two years on an international assignment from London. He has much experience in application design, development, and deployment.

He has worked on myriad technologies, including Visual Basic, SQL Server, SharePoint Technologies, .NET MVC, O365, Azure Data Factory, Azure Databricks, Azure Synapse Analytics, Azure Data Lake Storage Gen1/Gen2, Azure SQL Data Warehouse, Power BI, Spark SQL, Scala, Delta Lake, Azure Machine Learning, Azure Information Protection, Azure .NET SDK, Azure DevOps, and more.

He holds multiple Azure Certifications, including Microsoft Certified Azure Solutions Architect Expert, Microsoft Certified Azure Data Engineer Associate, Microsoft Certified Azure Data Scientist Associate, and Microsoft Certified Azure Data Analyst Associate.

Bhadresh has worked as solutions architect on large-scale Azure Data Lake implementation projects as well as data transformation projects, in addition to large-scale customized content management systems. He has also worked as technical reviewer for the book *Data Science Using Azure*, prior to authoring this book.

About the Technical Reviewer

Massimo Nardone has more than 25 years of experience in security, web/mobile development, cloud, and IT architecture. His true IT passions are security and Android. He has been programming and teaching how to program with Android, Perl, PHP, Java, VB, Python, C/C++, and MySQL for more than 20 years. He holds a Master of Science degree in computing science from the University of Salerno, Italy.

He has worked as a CISO, CSO, security executive, IoT executive, project manager, software engineer, research engineer, chief security architect, PCI/SCADA auditor, and senior lead IT security/cloud/SCADA architect for many years. Technical skills include security, Android, cloud, Java, MySQL, Drupal, Cobol, Perl, web and mobile development, MongoDB, D3, Joomla, Couchbase, C/C++, WebGL, Python, Pro Rails, Django CMS, Jekyll, Scratch, and more.

He worked as visiting lecturer and supervisor for exercises at the Networking Laboratory of the Helsinki University of Technology (Aalto University). He holds four international patents (PKI, SIP, SAML, and Proxy areas). He is currently working for Cognizant as head of cyber security and CISO to help both internally and externally with clients in areas of information and cyber security, like strategy, planning, processes, policies, procedures, governance, awareness, and so forth. In June 2017 he became a permanent member of the ISACA Finland Board.

Massimo has reviewed more than 45 IT books for different publishing companies and is the co-author of *Pro Spring Security, Securing Spring Framework 5 and Boot 2-based Java Applications* (Apress, 2019), *Beginning EJB in Java EE 8* (Apress, 2018), *Pro JPA 2 in Java EE 8* (Apress, 2018), and *Pro Android Games* (Apress, 2015).

Acknowledgments

I always believed that writing a book was a solo exercise, but after completing this book, which is also my first book, I realized that this is not entirely true. You will need support and motivation to continue writing the chapters one after another, tirelessly.

First, I would like to thank my wife, Priti, and my daughter, Jiya, for bearing with me over the many weekends and weekdays, for over six months, on which I devoted my time to this book after completing my day job. The time that was meant to be given to them, I have devoted to writing this book; hence, a very big thank you to both of them. Without their support and constant motivation to complete the chapters on time, it would have not been possible to complete the book.

There are a couple of friends and colleages whom I would like to thank for their support, guiadance, and motivation. I was not initially sure if I could author a book. So, I reached out to a couple of people who have authored books through LinkedIn before deciding to write this book. I would like to thank Mitesh Soni (`https://www.linkedin.com/in/mitesh-s-136842b`) for help and guidance on this matter. I would also like to thank my colleague and friend Amit Ubale (`https://www.linkedin.com/in/amit-ubale`) for the constant support and motivation throughout the entire book-writing exercise.

I got constant support from Apress throughout the entire book-writing process. Special thanks go to Smriti Srivastav, acquisition editor, for the initial contract and clarifying all my doubts promptly; Shrikant Vishwakarma, coordinating editor, for genuine follow-ups for each chapter, which kept me on track; and Laura Berendson, development editor, for reviewing each chapter minutely. Finally, I thank Massimo Nardone for his technical review of my book.

Introduction

This book is about beginning to learn Azure Synapse Analytics. It is the most modern data analytics platform offered by Microsoft. It is always challenging to start learning a new IT skill, as you will have to learn new jargon, concepts, methods, tools, and technologies as part of the process. Azure Synapse Anlaytics is no different. I hope that this book will give you a solid technical foundation for starting your journey in the new world of Azure Synapse Analytics.

Anyone with a little IT background, particularly in Azure cloud, will be able to understand the content of the book very easily. The book is targeted to data analysts, data engineers, data scientists, data architects, solution architects, Azure developers, Azure administrators, and so forth who are keen to begin their Azure Synapse Analytics journey. This book will be a good starting point. I have concentrated on some basic concepts in the first two chapters so that it becomes easy for anyone to start learning about Azure Synapse Analytics.

The book has a total of ten chapters, and I recommend that you read them in sequence if you are new to data analytics. If you have a basic understanding of data analytics, data lake, data warehouse, data lakehouse, etc., then you can skip the first two chapters and jump to the third chapter directly. If anyone is interested in any specific components of Azure Synapse Analytics, then one can jump to that component's specific chapter directly as well. Here I have given a very high-level summary of each chapter in a very brief manner:

> Chapter 1: "Core Data and Analytics Concepts" introduces readers to some of the important core data and analytics concepts as a foundation.
>
> Chapter 2: "Modern Data Warehouse and Data Lakehouse" provides conceptual understanding about traditional/legacy data warehouse, modern data warehouse, and finally the most modern data lakehouse.

Chapter 3: "Introduction to Azure Synapse Analytics" builds foundational knowledge by introducing Azure Synapse Analytics, its main features, and its key services capabilities.

Chapter 4: "Architecture and Its Main Components" explains Azure Synapse Analytics' core architecture and its main components, as it is very different from traditional data warehouse architecture and its components.

Chapter 5: "Synapse SQL" explores Synapse SQL in detail, including its architecture, its main features, and some how-to's to make readers familiar with some important activities that can be carried out for Synapse SQL.

Chapter 6: "Synapse Spark" explains Synapse Spark, its main components, and Delta Lake, along with some how-to's to make readers familiar with important tasks pertaining to Synapse Spark.

Chapter 7: "Synapse Pipelines" introduces Azure Synapse Pipelines and explans various types of pipeline activities with examples.

Chapter 8: "Synapse Workspace and Synapse Studio" familiarizes readers with Synapse Workspace and Synapse Studio, including its main features and its capabilities and how to accomplish some important tasks.

Chapter 9: "Synapse Link" explains the differences between OLTP and OLAP, why HTAP is required and its benefits, and then introduces Synapse Link along with its Cosmos DB integration, its features, and some use cases.

Chapter 10: "Azure Synapse Use Cases and Reference Architectures" discusses some of the industry use cases and reference architectures to introduce readers to real-life usage of Azure Synapse Analytics.

Happy reading!

CHAPTER 1

Core Data and Analytics Concepts

A few years back, oil companies used to rule the business world. That is no longer the case. Now data is considered the new oil, and there are multiple reasons to agree with this idea. Due to the explosion in social media platforms, a high volume of data is generated on a daily basis. Additionally, **Internet of Things (IoT)** devices generate a significant volume of data. Similarly, a variety of data is being generated and stored at a never-before-seen high velocity. As a result, organizations of all sizes across the globe have started to utilize the available data to their advantage. In today's modern era, each organization strives to become a data-driven one by introducing data-driven decision-making processes so as to stay ahead of competition in the market.

The IT industry has adapted to the changes in the data landscape very quickly. Many changes have happened rapidly regarding the kind of work we do in IT, as well as the names of the roles. For example, data analyst, data engineer, data scientist, and so forth are some of the new roles that have gotten popular in recent times. We also started to deal with all types of data, including structured data, semi-structured data, and unstructured data, along with the different ways in which data is made available for processing, including batch and streaming data. You may be familiar with the basic concept of databases, but let us start with core data concepts as a refresher.

Core Data Concepts

Before we jump into data analytics in detail, let us go through some of the core data concepts. If you feel that some or all of the concepts are known to you, then please feel free to jump to the next section or chapter. Here, we will go through some of the basic concepts, like data, structured data, unstructured data, semi-structured data, batch data processing, and streaming or real-time (RT) data processing.

1

© Bhadresh Shiyal 2021
B. Shiyal, *Beginning Azure Synapse Analytics*, https://doi.org/10.1007/978-1-4842-7061-5_1

What Is Data?

Let us look at the definition of "data" before we dive into other core data concepts. The word "data" is a plural of the word "datum." The word "datum" is originally from Latin. It means a piece of information. Thus, data are multiple pieces of information put together. Basically, we are talking about a bunch of information when we talk about data. In today's era of rapidly advancing technology, the word "data" has become very common in usage. "Data science" and "data engineering" have become very popular terms, which has increased the importance of data as well.

In the IT field, we know that any software is divided into two parts. One part is the code of the software, and the other part is the data on which the software code will take some actions. There are various types of data with which we deal regularly. Here, we are not talking about data types in a programing language paradigm. Generally, based on how the data is arranged or stored, we can divide data into the following three categories or types: structured data, semi-structured data, and unstructured data. Let us try to understand each one of them.

Structured Data

Structured data is any data that is highly organized and conforms to a specific schema or format. Each data point is placed in a highly organized manner so that it becomes very easy to interpret for both human beings and machines like computers. Since the beginning of the IT era, computers have been highly efficient in understanding and processing structured data. Data that we store in spreadsheet programs like Excel, as well as any data stored in database systems like SQL Server or Azure SQL Database, are basically structured data stored in a tabular format.

Let us look at an example of structured data. Let us assume that you have been asked to create an Excel sheet containing details of all the employees of your organization or of a specific department within your organization. You enter column names like Employee ID, First Name, Last Name, Date of Joining, Date of Birth, and Salary. You also enter details about a few employees. We can say that the data stored in your Excel sheet is structured data because it is highly organized and all records conform to the schema (list of all those columns you defined). Similarly, if you decide to store employee details from that Excel sheet in a database like SQL Server by creating a new table in it, then that would also be considered structured data.

Semi-structured Data

Semi-structured data is any data that is *not* fully organized but still partially conforms to a specific schema or format. Data stored in JSON or **J**ava**S**cript **O**bject **N**otation **(JSON)** documents and XML or e**X**tensible **M**ark-up **L**anguage **(XML)** files are good examples of semi-structured data. Instead of field names or columns, which we use for structured data, here we use tags to define different data points within a record. It is not considered highly organized since the order of the columns as well as the number of columns for each record may vary based on actual requirements. It means that semi-structured data lacks a fixed or rigid schema or data model. Generally, data and schemas are mixed without separation. Key–value stores and graph databases are used to store semi-structured data.

Let us assume that your IT team has given you a JSON file that contains your employees' details. If you open the file and read the data inside it, you will notice that there are tags for each data value. You may also find that for a few employees there are additional tags that are not available for all employees. For example, for a few employees, it may show you a nested structure to store previous roles or positions that they held in the past in your organization. This may not be present for those employees who have not held more than one role or position. Here, the employee data is somewhat organized and partially conforms to some schema or data model, so it is called semi-structured data.

Unstructured Data

This is the most important type of data that you should know about in more detail as around 85 percent of corporate data in this world is unstructured data. Any data that is *not* highly organized and *does not* conform to a specific schema or format is called unstructured data. In the last few years, it has gotten increasing attention due to its volume and the value it brings to data-driven decision-making processes. Businesses face huge challenges in storing, processing, and extracting hidden information from unstructured data.

Documents, image files, audio files, video files, and so forth are all examples of unstructured data. It is very difficult to gain meaningful insights from unstructured data as it is not highly organized, and it does not have a specific schema either. For unstructured data, you may try to store some properties at the file level that allow you to store additional information about the data being stored inside the file containing unstructured data. However, these properties will not be able to give you all the details

3

about unstructured data. For example, if you have to process a lot of image files, there is no simple way to do so apart from using some advanced machine learning algorithms that can scan through each image and predict what is inside each image.

Storage of unstructured data was also a real challenge as we cannot store such data in a database since it does not conform to any schema. In the last few years, cloud-based blob storage offerings from various public cloud companies have solved this problem to a large extent. Azure has multiple options to store your unstructured data. Azure File Storage, Azure Blob Storage, Azure Data Lake Storage Gen 1, and Azure Data Lake Storage Gen 2 are various types of storage options available to us to store unstructured data in the Azure cloud. Generally, it is preferred to store it on cloud storage, which may provide highly secured unlimited storage capacity and greater scalability with various cost-effective options that you can choose based on your actual storage requirements.

We have looked at various types of data based on how it is organized and stored. Now, let us see different ways of processing the data.

Data Processing Methods

The data that you receive from source systems will be in raw and unprocessed form. To derive meaningful insights from raw data, it needs to be processed first. This process of converting and transforming raw data into a usable and desired form is known as data processing. There are different methods of data processing based on how and when you process the data.

Batch Data Processing

This is the most common and traditional way of data processing. When you collect the newly ingested data into a group before processing it all together, it is known as batch data processing. Data ingestion may happen multiple times before you process it as part of a single batch at a future time. There are multiple pros and cons for batch data processing. Based on your requirements, you can decide if batch data processing will be suitable or not.

It allows you to process the data during non-business hours. It also provides flexibility to process a large volume of data at once. Many large data processing jobs are executed during night hours, when it will have the least impact on the business. Generally, data collected from various **on-line transaction processing (OLTP)** systems during business hours will be processed when business hours are over as part of overnight batch data processing jobs. Based on business requirements, you may also decide to run batch data processing jobs more than once during the day. For example, instead of processing all data together during overnight hours, you may decide to process data every four or eight hours.

By now, you will have noticed that batch data processing delays the availability of processed data, as the ingested data is not processed immediately but rather later, in a batch. There are many business scenarios in which this is not an acceptable approach for data processing. Another limitation of batch data processing that you need to consider is that all the required data should be available at the time of batch data processing, without any errors in it. Erroneous data or incomplete data will result in a delay in the batch data processing schedule, which may impact the business outcomes.

Let us assume that you download sales data for all salespersons at the end of each day from your sales transaction system. Based on that, you run a job to update the consolidated sales report containing the sales details of each salesperson for each day. This is an example of batch data processing in which you collect and group the ingested data for a day in a batch and then process it to update the consolidated sales report for that day.

Streaming or Real-Time Data Processing

If each new piece of data being ingested is also immediately being processed, then it is called streaming data processing. It is also known as real-time (**RT**) data processing, as it processes the data in real time without any major delay. Unlike batch data processing, there is no waiting after the data is ingested, as the data processing takes place in real time with a delay of only a few seconds or milliseconds. There are business scenarios in which new and dynamic data is generated continuously. These scenarios are the most appropriate in which to use real-time data processing or streaming data processing. Time-critical business applications will require streaming data processing, as any delay of a few seconds or in certain situations a few milliseconds will result in an adverse impact on business outcomes.

While batch data processing has all the data in a batch to be processed later, streaming data processing has only the most recent data or data generated within a rolling time window of less than one minute or so. Batch data processing can handle a very large volume of data in a batch, while streaming data processing processes only a few records at a time, as it must process the data in real time. These differences imply that you should use batch data processing for large and complex analytics workloads, while streaming data processing should be used for simple calculations or aggregations due to its very limited time window for data processing.

An **a**utomated **t**eller **m**achine (**ATM**) is a good example of the use of streaming data processing. The data processing happens in real time without much delay. It debits your account in real time, and the ATM dispatches the money instantaneously. Many financial institutions will track the stock market price movements in real time and update their risks immediately to avoid any losses. With streaming data processing, they can also provide real-time stock advice to their valuable customers. A lot of sensor data being used by autonomous vehicles (self-driving cars, etc.) requires streaming data processing, as any delay in processing may prove fatal. Industrial IoT devices and wearable smart devices like health bands, smart watches, and so forth use streaming data processing.

Let us assume that you are responsible for updating a sales report that includes the details of each sale made by each salesperson in real time. Here, it will become necessary for you to stream the data and process it in real time as and when a sale is registered in your sales transaction system. This will allow you to see the real-time sales numbers on your dashboard.

Relational Data and Its Characteristics

In bygone days, each application used to store data in its own unique data structure, which made it very difficult to make changes to it. This data structure was hard to design, code, and maintain. To solve this problem, **r**elational **d**ata**b**ase **m**anagement **s**ystems (**RDBMS**) were introduced. Relational databases made life easy for developers as they followed the specific pattern of a relational data model, which was easy to replicate across many applications. It was also very easy to design, code, and maintain such relational database management systems.

There are some important characteristics of relational data that we will go through one by one here:

- **Table:** A real-life object can be considered an entity for which we want to store information or data. This can be directly mapped to a table. Thus, a table is very basic but is the most important characteristic of relational data.

- **Rows and columns:** A table consists of one or multiple rows, which allow us to store multiple records or information about multiple similar objects or entities in a single table. Each row will have one or more columns to store different types of information about the real-life object or entity. Each column can store only one type of value for each row. Each row will have the same number and types of columns and in the same order as well.

- **Primary key:** To maintain uniqueness within a table or to identify each record within a table uniquely, at least one column across all rows must have different and unique values. That column is known as the primary key of the table. There may be a scenario in which you will have to combine more than one column to have uniqueness within the table. These columns are collectively called the composite key.

- **Relationship:** This is a very important characteristic from which the relational database derives its name, as it allows you to establish relationships among tables within the database. We can define a relationship between two tables by setting the primary key from one table to be a foreign key in another table. This gives us the ability to implement real-life objects' relationships very easily in relational data.

- **Foreign Key:** To create relationships between two tables, it is necessary that the first table's primary key is present in the second table, and that column in the second table must be defined as the foreign key. This way, a foreign key is the column, or the combination of columns, present in the second table, and the same column or combination of columns is defined as the primary key in the first table. This plays a crucial role in defining relationships between two tables.

- **Structured Query Language (SQL):** It is necessary to have the ability to easily query the stored data from relational tables. Apart from querying data, SQL also supports various other operations, like creating a table, inserting records into a table, updating records into a table, and deleting records from a table. It also allows you to filter rows based on where conditions, which you can specify in your queries. It also makes it possible to combine or join two or more tables together, allowing you to retrieve data from two or more tables at the same time using a single query statement.

- **ACID Properties:** ACID stands for **a**tomicity, **c**onsistency, **i**solation, and **d**urability. Any changes to the database are done via a transaction, which is a single logical unit of work to be carried out on the database. ACID properties are followed in most of the relational databases to maintain consistency before and after the transaction. Atomicity ensures that either the full transaction succeeds, or it fails. There is no partial success or failure allowed. Consistency ensures that the correctness of the database is maintained before and after each transaction. It enforces integrity constraints in the database. Isolation ensures that multiple transactions can occur simultaneously without compromising consistency in the database. The changes being carried out by one transaction will not be visible to the other transaction until the first transaction is committed successfully. Durability ensures that once the transactions are committed to the database, they are permanently stored in persistent storage, which can retrieved easily in the case of any system failure.

Azure provides multiple options for storing relational data in various RDBMSs. Azure SQL Database, Azure Database for PostgreSQL, Azure Database for MySQL, and Azure Database for MariaDB are various RDBMS options available on Azure for storing relational data. Each option has its own pros and cons, so it is necessary to explore each in detail to check which one meets most of your business and technical requirements.

Non-Relational Data and Its Characteristics

There are many real-life scenarios that will not have a relational data structure. For example, you cannot store video, audio, images, text data, and so forth in a relational database, as these are not relational data, but you still need to deal with these types of data. These are non-relational data, and you will have to store them in non-relational databases or repositories. These data will not have a relational structure, so they cannot be stored in a table in a relational database.

Generally, non-relational data is bigger in volume than relational data, and it also comes in a variety of data formats. It is necessary to understand the important characteristics of non-relational data. So, let us check a few of them out here:

- **Containers:** As we know, tables are used to store relational data. Here, we can use containers to store non-relational data. This allows us to store many entities or objects with different schema or different fields. Thus, containers support varied schema, which is not possible for a table in a relational database.

- **Storage Flexibility:** Non-relational data will not have a fixed schema. It is necessary to have storage flexibility to store data with varied formats, like audio, video, image, text, and so on. Non-relational data is stored in its native format, which is an important characteristic as it will allow you to deal with non-relational data in its raw format.

- **Data Retrieval:** For non-relational data, generally a unique key–value pair is used to identify an entity or object. Many non-relational databases support key–value pairs, which helps in retrieving non-relational data easily. It is recommended you use key–value pairs only to iterate through or filter the entities or objects stored in containers. Non-relational databases will provide their own SQL-like data-retrieval mechanisms or may have their own procedures for data retrieval.

- **Indexing:** This is not supported by all non-relational databases. This characteristic is similar to relational databases to a great extent. If the non-relational database you are using supports indexing, then your queries should be able to use the benefits of indexing to identify and fetch data for fields other than key–value pairs.

When you plan to use non-relational databases, it is very important to understand the preceding characteristics and check which of these are supported by your choice of non-relational database. Azure provides multiple options for storing various types of non-relational data. Azure Cosmos DB, Azure Blob Storage, Azure File Storage, and so forth are some of the non-relational data storage options provided by Azure.

Core Data Analytics Concepts

By now we have covered some of the important core data concepts, so now let us try to examine certain core data analytics concepts, which will give us a detailed understanding of some of the key processes or main activities that are performed in a specific sequence for any data analytics project.

What Is Data Analytics?

Due to the rise in importance of data in the business world, data analytics is considered as one of the most talked about areas in business and IT. In simple words, data analytics means the systematic and computational analysis of data. It is meant to find meaningful patterns in data, its interpretation, and its presentation or communication to relevant stakeholders. It helps organizations to make effective and efficient decisions.

Data analytics is different than data analysis, but you may find both terms being used interchangeably in IT literature. Data analysis is more focused toward the past and tries to find answers for what happened and why it happened. Data analytics is more focused on why it happened and what will happen in the future. This ability to predict what is going to happen in the future is the most important characteristic of data analytics for organizations. It allows organizations to use the insights received from data analytics to take appropriate actions or provide enough guidance for data-driven decision making.

Data analytics is a multi-disciplinary field. You need to apply knowledge and methods from statistics, computer science, and mathematics simultaneously to derive meaningful insights using data analytics.

Data Ingestion

The process of identifying and acquiring data sources and then importing the data from those sources is known as data ingestion. As part of data ingestion, raw data is imported from the source to the destination and stored in persistent storage for further data processing. Based on how you are receiving the source data—whether it is batch data or streaming data—you may ingest the data as a batch or a continuous stream.

The data ingestion process may also involve some data filtering activities. You may decide to reject corrupt or duplicate data as part of your data ingestion process. It may also involve data transformation activities, like changing the date format or converting amount fields to a specific currency from many different currencies. The important point to note here is that the transformations you want to apply at the time of data ingestion should be very lightweight so as to not impact the overall data ingestion timeline too much, as well as to not make big changes to the raw data. If you need to carry out complex transformations, then those should be carried out once the data is ingested and stored in persistent storage. Complex transformations should be part of the data processing phase and should not be part of the data ingestion process.

Data Exploration

Once the data ingestion is done, the raw data in its original format is available for data exploration purposes. Due to the rise of data science, **EDA**, which stands for **e**xploratory **d**ata **a**nalysis, has gotten a lot of momentum in the data field. Data scientists, data engineers, and data analysts need to explore the data in order to make certain important decisions about how it should be processed further. Data exploration may give insights about the data's formats, sizes, schemas, and more. These inputs are important for designing data processing pipelines. Data exploration also helps in designing an end-to-end machine learning pipeline.

Data professionals use various methods for data exploration. If the volume of data is huge, then it is very difficult to analyze the data in a tabular format. In this situation, data professionals use various data visualization techniques such as charts, graphs, and so forth to visualize the data. Data exploration also allows you to identify abnormal or outlier data points very easily. Data scientists use data exploration techniques via EDA to know the data better, which allows them to pick the most appropriate machine learning or deep learning algorithm for the given dataset.

Data exploration also helps in identifying any data cleansing requirements up front. You can check if your data can pass through basic validation checks or not. Data exploration can help you identify if there are any NULL values in those fields in which you are expecting a value. Similarly, it can also help you to identify whether there are duplicate records in the data. You can also check if the values from a particular field are within the acceptable range or acceptable list of values. Thus, data exploration can help you in multiple ways to make crucial decisions up front so as to avoid any rework or unnecessary work later in the development cycle.

Data Processing

Once you have identified, collected, ingested, and stored the data in persistent storage like Azure Data Lake Storage, you are ready to start data processing activities. Generally, the data processing stage will take the raw data in its original format, cleanse it, and transform it into more meaningful formats, like tables or documents. Data processing is likely to include many business transformation rules, using which multiple source objects or files will be joined to generate new target tables or files. The transformed data will be stored in persistent storage like Azure Data Lake Storage as a separate data layer.

The Raw Data layer or Raw landing zone is the main source of data for the data processing stage. The Curated Data layer is the main target of the processed data and is where the transformed objects or files are stored. Azure Data Lake Storage is generally used as both the Raw Data layer and the Curated Data layer. There are various terms used for Raw Data layer like Raw landing zone, Raw Data vault, Universal data lake, and so on. Similarly, the Curated Data layer is also known as the Processed Data layer, Curated Data vault, Business data lake, and so forth.

In data analytics, the data processing stage will generate analytical data models from which a business can derive more value, as the raw data is enriched as part of the data processing stage. In many cases, data processing is a very time-consuming and complex stage in the whole data analytics process. Due to its complex nature, it is highly recommended you automate the data processing stage by using various tools. Azure Functions, Azure Databricks, Azure Synapse Analytics, and so on are some of the well-known tools that are widely used for lightweight to heavy duty data processing purposes.

As part of the data processing stage, there are two different approaches to transforming or processing the data. If you transform the data before loading it to target storage, then it is known as ETL, and if you load the data first to target storage and then transform it, then it is known as ELT. Let us try to understand these two different approaches.

ETL

ETL stands for extract, transform, and load. Here, once you extract the data, you also apply various business transformation rules on it before you load it to the target; hence, it is known as the **ETL** or **e**xtract-**t**ransform-**l**oad approach of data processing. This is the traditional way of processing the data. It is still widely used in many scenarios. Generally, the three processes of extraction, transformation, and loading will be part of a continuous data pipeline. Since the transformation has to happen before loading the data to target storage, it is generally implemented as a continuous series of operations put together to extract, transform, and load the data in a single data pipeline. It should be used in only those scenarios where only basic business transformations and basic filtering are required. The ETL method is not recommended for those scenarios in which you need a complex business transformation that is going to be a very time-consuming process.

Let us assume that you have some requirements to apply some column-level business transformations and some row-level basic filtering while extracting and loading the data for Employee Master from a CSV file being received from your HR system. In this scenario, it will be suitable to use the ETL method, as lightweight transformations can easily be applied before loading the Employee Master data to target storage. As it allows you to filter the data up front before it gets loaded, it is also a very useful method in those scenarios in which you want to avoid loading everything to the target and want to discard the full records and/or certain columns as part of your data processing.

ELT

ELT stands for extract, load, and transform. Here, once you extract the data, you directly load the extracted data to the target, and then, after loading the data, you run separate data processing tasks to transform the data. Hence, it is known as the **ELT** or **e**xtract-**l**oad-**t**ransform approach of data processing. This is a newer approach than ETL for processing data. It is widely used in many scenarios. Once the extracted data is loaded into the target storage, it is again retrieved to apply transformations, and then the data is written back to the target storage.

You need to have two different data pipelines to complete data processing using the ELT method. The first data pipeline will only extract and load the data as it is in the target storage without any business transformations. The second pipeline will retrieve the stored data from the target storage and apply various complex business transformations and multi-level filtering to process the data, and then it will write it back to the target storage. Here, the second data pipeline will be more complex and time consuming compared to the first data pipeline. It will also provide opportunities to combine multiple tables or files to create new target tables or files to support the needs of the business.

Let us assume that you have some requirements to combine Employee Master data and Sales data, which you receive from your HR system and sales system, respectively. In this scenario, the ELT method will allow you to extract and load the data to the target system by using two different data pipelines. Once the data is loaded from both systems, you will have to run another data pipeline to combine the already-loaded Employee Master data and Sales data together from the target storage. Once you combine the data using joins, it will create a new table for you, which you can store in the target storage. This new table will give you more meaningful business insights compared to two different tables. Hence, the ELT method is widely used to support such complex business scenarios for data analytics.

ELT / ETL Tools

There are multiple tools that support ETL/ELT methods. Traditionally, **SQL S**erver **I**ntegration **S**ervices (**SSIS**) is widely used as an ETL/ELT tool. It comes as a separate service with Microsoft SQL Server. However, it is not supported on the cloud directly. So, if you are dealing with legacy on-premises data sources from which you are extracting data, then there is a high chance that you will be using SSIS as an ETL/ELT tool.

Due to cloud computing getting very popular in the last few years, even traditional tools are being offered in cloud flavor, with various new features, robust reliability, and availability. **A**zure **D**ata **F**actory (**ADF**) is the cloud avatar of SSIS. It supports a wide variety of data sources with a lot of new features compared to SSIS. Azure Data Factory provides a very easy-to-use **g**raphical **u**ser **i**nterface (**GUI**)–based approach to develop the data pipelines visually. It is a no-code or low-code data pipeline building tool. It is fully web based so there is no need to install anything locally on your machine. It allows you to build very complex data pipelines very easily. Azure Data Factory also offers options to create data flows using other Azure Data Services like Azure Databricks, Azure SQL Database, Azure HDInsight, and Azure Synapse Analytics. Azure Data Factory is tightly integrated with Azure Synapse Analytics as Synapse Pipelines.

Data Visualization

Once you have done data processing and data exploration as per your requirements, the next step in the data analytics process would be to show the generated output data to the relevant stakeholders in the form of a report. Generally, a report contains the data describing what has already happened in past along with its finer details in rows and columns. A report should be easy to interpret for its stakeholders and should be able to answer most of the questions of its users. Tabular reports were famous for achieving this in the past, and this continues to be the case even today in many situations. Microsoft **SQL Server Reporting Services (SSRS)** is one of the most widely used reporting tools that provides server-based reporting services. It is part of Microsoft SQL Server services. SSRS can be used to prepare and deliver a variety of interactive reports.

Reporting evolved into more sophisticated **business intelligence (BI)** tools. Business intelligence is different than reporting. The latter gives information about what has happened in the past, while the former gives more input around why certain things happened and may suggest corrective actions to improve an organization's performance. Business intelligence refers to the tools, technologies, applications, and methods for data collection, integration, analysis, and presentation of the organizational information. Most of the BI tools support interactive reporting, drill-through and drill-down reporting, slice and dice pivot table analysis, and data visualizations.

As the size of the data being analyzed and reported is increasing continuously, it has become difficult to convey the real information stored in the data. It is not humanly possible to interpret the large volume of data in a tabular report format. Here, data visualization tools come to your aid. Data visualization is the graphical representation of information and data. Generally, data visualization tools use graphical elements like charts, graphs, maps, and so forth to convey relevant information to stakeholders, which can be more easily interpreted than the tabular report format.

Microsoft offers Power BI as a data visualization tool; it is a collection of software services, apps, and connectors that work together to turn your unrelated sources of data into coherent, visually immersive, and interactive insights. It is tightly integrated with many Azure services for a seamless experience.

Data Analytics Categories

Data analytics includes a range of activities or tasks. Based on the activities or tasks involved in an analytics scenario, we can divide data analytics into different categories. Let us get a basic introduction to each of these categories to understand what activities or tasks each of the categories includes and what the goal is for each category.

Descriptive Analytics

Descriptive analytics is the most preliminary technique for data analytics. It does exactly what its name implies: it describes or summarizes the data being analyzed. It is meant to analyze the raw data and convert it into a format that can easily be interpreted by human beings. It helps to answer the question around what has happened in the past based on the available historical data.

Let us assume that your organization has set a specific sales target to be achieved within a year. You have been asked to create a KPI to measure the aggregated sales numbers each month and indicate how close or far your organization is to that sales target for the year on daily, weekly, or monthly frequencies. Here, **KPI** or **k**ey **p**erformance **i**ndicator is a critical or key indicator of progress toward the expected or set outcome. In this example, it is the total sales number that your organization wants to achieve by the end of the year. This KPI will describe how many sales have happened in a past period and compares that with the intended outcome. There could be many such KPIs, like return on investment, number of customers retained, employee churn rate, monthly hits on your organization's website, and so on, which can be described easily by using the descriptive analytics technique.

Diagnostic Analytics

Diagnostic analytics is an additional exercise on top of your descriptive analytics work. It allows you to get answers to questions for a specific event. Descriptive analytics describes what happened in the past based on available historical data, while diagnostic analytics diagnoses the reasons behind why an event happened in the past. It takes the outcome from descriptive analytics and digs deeper to find the cause or reason for whatever happened in the past based on data.

KPIs described by descriptive analytics are further scrutinized to find why a KPI performed better or worse. As part of the diagnostic analytics process, you will have to find the anomalies or irregularities in the data. As a next step, you will try to find and collect additional data that may be responsible for the anomalies or irregularities found in the first step. In the last step, you will try to establish relationships or to find historical trends that may give a better diagnosis about why those anomalies or irregularities happened.

Let us assume that the sales target KPI described in the previous section shows abnormal growth for a specific month within a year. To know exactly why it happened, you will use diagnostic analytics. You will try to see the data collected for that specific month and analyze it to identify anomalies. You may find that a specific region that used to give good sales numbers each month has not performed well that month. So, to find why it is so for that region, you will dig deeper into the region-specific data for that month. You may find that there were a couple more holidays in that region compared to other regions, and that a few of the salespersons were on leave due to a regional festival, which resulted in lesser than expected sales in that region for that month.

Predictive Analytics

In the previous two sections, we saw how descriptive analytics describes what happened in past and diagnostic analytics diagnoses the reason for that based on historical data. The next obvious step would be to know what will happen in the future. Predictive analytics helps in answering that question. It uses historical data to find trends and patterns in the existing data and determines if those trends and patterns will recur in the future. Generally, various statistical and machine learning–based techniques are used for predictive analytics.

Let us assume that you have collected sales numbers for the last ten months and you have been asked to come up with predicted sales numbers for the entire year. Here, you may rely on predictive analytics techniques and may use any statistical or machine learning–based method to predict the sales numbers for the remaining two months, which will also allow you to derive consolidated sales numbers for the entire year. It will show you clearly if your organization will be able to meet its target or not and if there is any shortfall or surplus then how much that shortfall or surplus will be at the end of the year.

Prescriptive Analytics

Data-driven decisions can be made by using insights from predictive analytics, as it will tell you if you are going to meet your goal or not. If you are not going to meet your goal, then prescriptive analytics can come to your rescue. It helps to answer questions about what actions should be taken to achieve your goal, which otherwise you are not going to meet. Organizations can use prescriptive analytics to make informed decisions based on facts presented by data. Primarily, it relies on machine learning techniques to find patterns. It allows you to analyze past events, actions, and so forth to estimate the likelihood of various end results based on patterns and trends.

What will you do if your organization asks you to recommend corrective action items to help your organization to achieve the sales target? You may rely on prescriptive analytics in the given scenario to give you some corrective actions based on the available historical data. Prescriptive analytics can tell you which parameters you will have to tweak to get the desired outcome and how you should be tweaking those parameters. In our case, it may tell you the desired number of working salespersons for the remaining period as well as indicate if you need to make any changes to your pricing policy for the products or services your organization is selling.

Cognitive Analytics

Cognitive analytics aims to mimic human intelligence. It will tell you what might happen if the given circumstances change and how you can handle emerging situations better, like a human expert giving you advice. It is one of the most promising analytics techniques and is advancing rapidly due to the world-wide research happening in machine learning, deep learning, and natural language processing (**NLP**). It uses available data and its patterns to generate inferences, uses existing knowledge to derive decisions, and finally generates a feedback loop to add all these findings back into the knowledge base for future inferences.

The most important element in cognitive analytics is its ability to extract and use knowledge from non-conventional sources like unstructured data stored in text files, image files, audio files, video files, and so on. Let us assume that for the tracking of a sales target for the year, you are using cognitive analytics. That means that you will have to feed additional inputs to the system, which will make intelligent decisions based on emerging situations and will provide real-time advice to you about how to tackle the specific situation that has just emerged.

Summary

Now you know some of the important Core concepts around data and data analytics. We started with core data concepts, which include a basic definition of data; various types of data, like structured data, semi-structured data, and unstructured data; and data processing methods like batch data processing and real-time data processing. We also looked at relational data and its characteristics, along with non-relational data and its characteristics at a very high level. Later, we moved on to core data analytics concepts, which include the basic definition of data analytics and its various stages, like data ingestion, data processing, data visualization, and so on. Toward the end, we also covered various categories of data analytics.

This chapter contains some foundational knowledge that will help you to understand the next chapter, as it is going to cover modern data warehouse and data lakehouse. That will give you a strong foundation to understand Azure Synapse Analytics and its related concepts in the subsequent chapters. See you in the next chapter! Happy learning!

CHAPTER 2

Modern Data Warehouses and Data Lakehouses

Databases are meant to store structured data, and so are data warehouses; however, data warehouses are different in many respects. Due to the increase in the number of applications used per organization, the increase in the volume of data, and the increase in the speed at which these data are generated, a specialized system is warranted that allows for the processing and aggregating of large volumes of data received from disparate source systems. That is where the data warehouse comes into the picture. There are yet other reasons why today's organizations need a data warehouse as part of their data analytics strategy. In this chapter, we will examine what a data warehouse is and why it is necessary for an organization to have one. As with all other IT systems, the data warehouse has evolved over time, and there are some significant improvements in its technologies that differentiate a traditional data warehouse from a modern data warehouse. We will look at these differences and improvements in detail in this chapter.

In last decade or so, Big Data analytics has achieved significant importance in both business and IT as it supports data analytics workloads for a large volume of data of various formats and velocity. So far, these Big Data analytics systems have remained separate from data warehouse systems. However, there is an increasing demand in the market to join these two technologies together. Nowadays, Big Data analytics systems rely on a data lake as a storage mechanism, as it allows one to store data in structured, semi-structured, and unstructured formats. The amalgamation of Big Data analytics using data lake and modern data warehouse technologies is known as a data lakehouse. It is a relatively new concept that is emerging quickly in the market. Databricks and Microsoft have become pioneers in advancing data lakehouse technologies in the market by offering various products in this area. We will cover several important topics in this chapter, like what a data lakehouse is, how it is different from a data warehouse, and the benefits it gives over data warehouses.

© Bhadresh Shiyal 2021
B. Shiyal, *Beginning Azure Synapse Analytics*, https://doi.org/10.1007/978-1-4842-7061-5_2

Before we get into data lakehouses, it is necessary to understand data warehouses. If you have foundational knowledge of data warehouses, then please feel free to skip these topics and jump straight to the data lakehouse topics. So, here we go!

What Is a Data Warehouse?

Traditionally, each business entity will have one or more business areas with different business models that consist of multiple simple to complex business processes. These business processes are supported by various business applications. These applications generate a large volume of data, which is generally stored in various databases depending on the nature of the business applications. Since these databases support business transactions during the course of the business, they are known as transaction databases, and these business applications are known as **t**ransaction **p**rocessing **s**ystems (**TPS**). These are also known as **o**n-**l**ine **t**ransaction **p**rocessing (**OLTP**) systems. These databases are continuously being used by business applications. So, if you try to use the same database for any business intelligence or analytics purposes to support your business decision-making process, you are bound to face issues. That is why business **d**ecisions **s**upport **s**ystems (**DSS**) are kept separate from business transaction processing systems. It is recommended that business intelligence or analytics workloads are run on their own database that can support complex and long-running queries to aggregate and summarize the business data captured by the TPS. This will also ensure that you do not interfere with transaction databases, which are being used by business applications to insert, update, and delete various data into the database during the course of the business' activities.

In bygone days, there were multiple decision support systems to support various business functions internally. Most of the time, the same transaction system data got transferred to all the decision support systems, as they all needed common data to support their function- or department-specific decision-making processes. This used to create multiple copies of the same data in different systems within the same business entity. This setup became chaotic and time consuming to build and maintain, as the number of business applications and the complexity of the same kept on increasing, along with the data volume. There was a pressing need to combine these decision support systems and give a single version of the truth to make the decision-making process more consistent and robust within a business entity. That is where the concept of the data warehouse came into the picture.

A data warehouse allows you to segregate your transaction systems and analytical systems. As its name suggests, it is a warehouse for data. This also implies that it will store historical data for transactions as well as reference data or dimension data to support various data analytics and machine learning workloads. Historical data can help a machine learning algorithm to predict the future and may allow businesses to make data-driven decisions to stay ahead of the competition.

With the preceding background in mind, let us try to define *data warehouse* in simple terms.

A data warehouse is a central repository of current as well as historical data that are collected and processed from one or more disparate source systems for reporting, business intelligence (BI), and data analytics purposes to support the data-driven decision-making process within an organization.

Core Data Warehouse Concepts

There are some important data warehouse concepts that you should understand clearly. These concepts are useful for understanding the purpose, structure, and high-level architecture of data warehouses. These concepts are the pillars of data warehouse technology.

Data Model

A data model organizes various elements of data, provides structure to it, defines relationships among various elements of data, and defines its relationship with real-world entities. Without a well-defined data model, it is very difficult to derive any meaningful or useful insights from the data stored in a data warehouse.

Generally, designing a data model for a data warehouse involves multiple steps. A conceptual data model is designed first at the object level without having any attribute-level details in it. Business-specific concepts are captured in the conceptual data model at a high level. Based on that, a logical data model is designed that will have attribute-level details as well as clearly defined object-level relationships. That means that you will have to define primary keys and foreign keys in your logical data model. At last, a physical data model is designed. Here, you will have to design the table structure with column names, column data types, primary keys, foreign keys, column constraints, and so forth. This will provide a clear view of how physically the tables will be stored and a specific table's relationship with other tables inside the data warehouse.

Model Types

A data warehouse can support two different types of models: relational data models and dimensional data models. Generally, a relational data model is used for OLTP systems or transaction processing systems. However, it is also possible to use it in a data warehouse. A relational data model tries to identify fundamental and important entities involved in business transactions.

On the other hand, a dimensional data model, which is widely used for OLAP systems like data warehouses, tries to identify entities that can provide business metrics like marketing metrics, sales metrics, financial metrics, etc. The dimensional data model derives its name from the fact that it allows you to view the data from multiple dimensions. Fact tables and dimension tables are important concepts to understand for the dimensional data model. Fact tables contain metrics, measurements, or facts about a business transaction, while dimension tables contain various dimensions of a fact. Basically, dimension tables are joined with fact tables via foreign keys. The joint view from a fact table and various dimension tables can provide a multidimensional view of the data. That view helps businesses to derive meaningful and useful insights from the data stored in a data warehouse using the dimensional data model.

Schema Types

Within a dimensional data model, there are a few different options for deciding the schema type. Generally, a star schema or a snowflake schema will be implemented for a dimensional data model in a data warehouse. A star schema is easy and simple to understand and implement. There will be a fact table in the center, and it will be surrounded by multiple dimension tables. Since it creates a star-like design, it is called a star schema. Here, fact tables are joined to various dimension tables using primary keys and foreign keys.

In the case of the snowflake schema, there will still be a fact table at the center, but it will be surrounded by not only dimension tables but also sub-dimension tables. This creates a design like a snowflake. Snowflake schemas are complex to understand and implement in a data warehouse.

Metadata

In simple words, metadata is data about data. Metadata describes or explains data. If you are going to receive data from many different source systems, then each source system may have different metadata to represent the data. So, it becomes very important to manage the metadata systematically. Metadata acts as a data dictionary of your data warehouse. It will allow you to search for and locate your relevant data easily and efficiently. Thus, it is important to have robust metadata management practices in place for a successful data warehouse implementation, as metadata is used by many tools that you will use in your data warehouse. These tools include querying tools, reporting tools, extraction tools, integration tools, data quality tools, transformation tools, and more.

Why Do We Need a Data Warehouse?

As mentioned at the start of the chapter, if we already have databases to store structured data, why do we need a separate data warehouse? Although a data warehouse is a relational database at the end, there are many differences between the two. Based on the definition that you saw earlier, you may have some idea about why we need a data warehouse, but let us explore this in a little more detail here.

Efficient Decision-Making

The most important reason to have a data warehouse is to support a more efficient and faster data-driven decision-making process for a business entity. It increases the ability to make faster decisions as the required data is already aggregated and summarized and ready for consumption for this purpose. If you don't have a data warehouse, you will have to fetch the data from the TPS and apply aggregations and transformations on it, which is going to take some time. The same information is readily available in a data warehouse, as you would have done that processing up front as part of your ELT or ETL processes.

Separation of Concerns

A data warehouse provides you with a separation of concerns, as you have a separate data warehouse that has an independent and separate TPS. This is necessary to avoid having bottlenecks at your TPS, as data warehouse will not interfere with while data is being inserted, updated, and deleted in the TPS. Typically, the data warehouse is read-intensive, while the transaction processing system is write-intensive, as you continuously post various business transactions into it. Hence, having a separate data warehouse from your TPS keeps your read-intensive and write-intensive workloads on two different systems, and in that way separates the concerns of normal business operations from the data analytics workloads required for business decision-making processes.

Single Version of the Truth

Assume that an organization has more than 50 small to large business applications to support various business-specific requirements, and around ten decision support systems to support the function- or department-specific decision-making processes. Here, you will end up sending some common data to all ten decision support systems. This will create multiple copies of the same data in multiple systems and will result in inconsistent data sources. So, it may happen that the same data points will give you different results if you query the same data from different decision support systems. This means there is no single version of the truth within your organization for a given data point. A data warehouse, however, provides a single version of the truth for your data and is preferred over disparate systems for data analytics–related workloads.

Data Restructuring

Changing the structure of the data is called Data Restructuring. A data warehouse creates an opportunity to restructure the data being generated in business applications. If table names and their column names are not very business friendly in transaction databases, then those can be renamed appropriately to make them very business friendly for the end users. Apart from that, a data warehouse also allows for opportunities to apply appropriate formats to the data. For example, a table named Emp001 can be renamed Employee_Master to make it more meaningful and interpretable. A column having a system-specific date format can be changed to make it more easily readable by end users.

Self-Service BI

Self-Service BI is an approach which enables business users to access and explore the data without much BI knowledge and expertise. A data warehouse can also support self-service BI to a large extent. It becomes easier for business users to generate reports on their own without intervention from IT staff as restructured data can be easily understood and interpreted by business users via the data warehouse. There are many BI tools that support direct connections to most of the well-known and widely used data warehouses. This helps increase data literacy within the organization. This is possible because data warehouses provide the opportunity to restructure data and give user-friendly names and formats to various objects.

Historical Data

A data warehouse can store a ton of historical data. It can easily handle a large volume of data, compared to traditional databases. Sometimes, it becomes difficult to maintain historical data in transaction databases due to various system limitations as well as cost escalations. As a result of various regulatory and legal requirements, one might have to retain historical data for a longer duration so as to support auditing of the legacy data. A data warehouse can support the retention of a large volume of data in a cost-effective manner; this data can be queried when required. It is also very important to note here that keeping historical data within transaction systems may impact the systems' performance as the data volume keeps increasing each day. Hence, it makes perfect sense to move historical data to a data warehouse to lessen the burden on the transaction systems.

Security

As a data warehouse is a central repository of current and historical data, you can provide access to the data and manage it centrally and securely for various types of users based on their requirements. You can have a common security model for accessing the data stored in a data warehouse, as opposed to managing direct access to multiple transaction systems via different security models or access methods. Data warehouses provide various options for authentication and authorization so as to give secure access to data.

Data Quality

The quality of data has become increasingly important as we continue to generate a large volume of data. However, if the data quality is poor then it will adversely impact the decision-making process of the organization. So, it is very important to maintain a high level of data quality to enable making the correct decisions efficiently.

When you integrate data from various source systems into a data warehouse, you have an opportunity to clean the data by applying appropriate data quality checks. This helps to maintain a high level of data quality in the data warehouse, which in turn results in quality decisions' being made based on that data.

Data Mining

A data warehouse is the right place to apply various data mining techniques by using myriad data mining tools. Data mining can easily unearth hidden patterns and trends in the data stored inside your data warehouse. This is very useful for businesses, as it can tell them where their business is headed in the future.

More Revenues

Those organizations that have invested in and implemented a data warehouse have generated more revenue from their respective businesses than those that have not invested in and implemented a data warehouse. A data warehouse allows key decision makers to make correct data-driven decisions very quickly and efficiently, which an organization without a proper data warehouse will struggle to achieve. That is the main reason why an organization with a data warehouse can generate more revenue.

What Is a Modern Data Warehouse?

In Chapter 1, we discussed relational and non-relational data as well as their characteristics. A modern data warehouse consists of both relational and non-relational data. Non-relational data includes both semi-structured data and unstructured data. The volume, variety, and velocity of data have increased drastically due to the explosion in data related to social media and Internet of Things (IoT). Nowadays, more data is generated by IoT devices than is generated by human beings. Thus, the traditional data warehouse has evolved into a modern data warehouse that supports all types of data.

A data warehouse which supports all types of data and can provide faster response time is considered as a modern data warehouse. Generally, a modern data warehouse is built on public clouds like Azure, AWS, Snowflake, GCP, and so on. Due to the inherent nature of cloud technologies to provide separation of concerns for various cloud reesources like compute, storage, network, etc., a modern data warehouse hosted on the cloud provides a clear separation of compute and storage resources. That results in cost savings and increased flexibility for a modern data warehouse compared to a traditional data warehouse. There is no need to invest upfront in infrastructure when you host your data warehousing projects on the cloud; rather, you will be provisioning your infrastructure instantly on your choice of public cloud. Cloud platforms also provide the option to scale the data warehouse automatically as and when there is a need to do so, with some minor restrictions. Modern data warehouses allow you to scale compute and storage resources independent of each other, unlike a traditional data warehouse.

Modern data warehouse architecture includes ELT/ETL tools like Azure Data Factory, data processing tools like Azure Databricks, data lake storage like Azure data lake Storage, relational databases like Azure SQL, analysis services like Azure Analysis Service, and data visualization and analytics tools like Power BI, Tableau, and so forth.

Difference Between Traditional & Modern Data Warehouses

Traditional data warehouses have evolved into modern data warehouses with some commonalities like disparate source systems, data models, schematic layer, consumption layer, etc. However, there are some noteworthy differences between them. Let us discuss some of them.

Cloud vs. On-Premises

Generally, a traditional data warehouse is built using on-premises servers hosted on business-owned or rented data centers. That means that if you have to build your data warehouse from scratch, you will have to invest in physical infrastructure up front. A modern data warehouse is exactly the opposite in this aspect, as it is built on public clouds like Azure, AWS, Snowflake, GCP, and so forth.

Separation of Compute and Storage Resources

Compute resources and storage resources in traditional data warehouses are generally not separable. Due to the inherent nature of cloud technologies to provide separation of concerns, modern data warehouses hosted in the cloud provide a clear separation of compute and storage resources. That results in cost savings as well as increased flexibility, as you can increase the capacity of your compute resources without impacting your storage resources and vice versa.

Cost

There is no need to invest upfront in physical infrastructure when you start your data warehousing projects hosted in the cloud; rather, you will be provisioning your infrastructure instantly on your choice of public cloud. Here, you may incur operational costs over the period of your data warehouse usage, and will have the option to select a suitable **Pay-As-You-Go** (**PAYG**) payment plan. So, your initial cost will be much less as compared to traditional data warehouses, for which you would have to invest in infrastructure up front.

Operational costs for a traditional data warehouse, however, will be less compared to those for a modern data warehouse that is hosted in the cloud. This is an important difference to know between a traditional and a modern data warehouse. Generally, data warehouse projects are considered costly propositions, but modern data warehouses based on cloud technologies have eased this burden heavily, as you need to worry only about ongoing operational costs and not the upfront capital costs required for traditional data warehouses.

Scalability

If you want to increase the capacity of your traditional data warehouse to meet additional workloads, you will have to order and procure new physical infrastructure. This option will further be hampered by the long procurement cycle, and, once procured, it will take time to set up and configure the procured infrastructure so that it can be utilized for the data warehouse.

A cloud-based modern data warehouse provides options to instantly scale the data warehouse based on your requirements. If there is a temporary spike in usage, to accommodate that surge in demand you need to temporarily scale your data warehouse. This is not possible in the case of traditional data warehouses unless you have invested in spare infrastructure capacity. Modern data warehouses can support this scenario easily. You can increase your resources during that brief period to accommodate the sudden surge, and once the usage comes down to a normal level, you can reduce your resources to a normal level.

ETL vs. ELT

As part of Chapter 1, we discussed ETL and ELT processes and what the differences are between the two. It is common to have an ETL process for loading data into traditional data warehouses. This means that those who want to analyze data from a traditional data warehouse will have to wait for the ETL processes to extract, transform, and load the data. This will increase the overall time needed to generate the insight and intelligence from the data. Meanwhile, a modern data warehouse uses an ELT process, which extracts and loads the data faster than ETL, as the loaded data will be transformed later. This gives earlier accees of source data to data analysts, data engineers, data scientists, etc. than an ETL process would give. It also gives access to source data in its raw or original formats, which is generally a requirement for your data engineering and data science teams.

Disaster Recovery

In the case of a traditional data warehouse, a lot of extra effort is needed to set up a disaster recovery site. You will have to identify the site and invest in infrastructure up front. A modern data warehouse built on a cloud platform will have many cost-effective and efficient options for setting up and configuring your disaster recovery site. It will be less time-consuming and easier to set up the disaster recovery site on a cloud platform compared to achieving the same objective using on-premises infrastructure.

Overall Architecture

There are major differences between the overall architecture of a modern data warehouse and that of a traditional data warehouse. Traditional data warehouses follow a three-tier architectural approach, which includes a bottom tier, middle tier, and top tier. The bottom tier consists of a relational database on which to store data, which is loaded from various source systems after cleansing and transforming it. The middle tier consists of an OLAP server, which works as a mediator between the bottom tier and top tier. Generally, a relational OLAP or a multi-dimensional OLAP model is implemented as an abstraction layer in the middle tier. The top tier is where the end users interact with the data warehouse via various front-end tools and techniques. Myriad querying, reporting, BI, and mining tools and techniques are used in the top tier to interact with the traditional data warehouse.

Compared to this, a modern data warehouse's architecture is quite different. While it maintains some of the same core components, many new components were introduced. Generally, it will include a data lake into which data will be loaded from disparate data sources after extraction. To transform the data loaded into the data lake, a data processing engine like Databricks will be used; it will then load the data into a relational database, from which the data will be consumed by end users directly or through various reporting, BI, analytics, and visualization tools.

These are not the only differences between a traditional data warehouse and a modern data warehouse. For example, there are differences in how you could extend your traditional and modern data warehouses. Similarly, if you want to make your data warehouse highly available to meet the Service Level Agreements (SLAs) expected by a business, each type of warehouse has different ways to achieve this.

Data Lakehouse

In the last few years, a trend has emerged in which there is an increasing need to combine data warehouse capabilities with data lake capabilities. These are different concepts but, due to the needs of the market, currently data warehouses and data lakes are in convergence. There are immense benefits for the data landscape that this convergence is bringing to the table.

In previous sections, you have learned what a data warehouse is, what a modern data warehouse is, how it is different from a traditional data warehouse, and so on. However, so far we have not discussed data lakes, Delta Lake, Apache Spark, and so forth. It is very important to understand certain basic concepts around these components, as that will help you to understand the concepts around data lakehouses in this chapter and Azure Synapse Analytics later in this book. So let us start with a general question: What is a data lake?

What Is a Data Lake?

Around a decade ago, non-relational data exploded fully, and there were no simple and easy-to-use systems in the market to handle that. Databases are meant to handle relational data, which are fully structured in format and comply with a specific schema. Non-relational data consists of semi-structured data and unstructured data that cannot be handled efficiently using a database system. Big Data emerged as a new trend as the volume, variety, and velocity (**3Vs** of Big Data) of data increased drastically. Hadoop was the option available to handle that situation at that time. It uses the **Hadoop Distributed File System (HDFS)** to store all types of data, including structured data, semi-structured data, and unstructured data. So, in simple words, we can say that any system that supports a file system like HDFS or allows you to store objects like BLOBs (**B**inary **L**arge **Ob**jects) is known as a data lake. Another important point to note is that generally you store all types of data in a data lake in the original or raw format in which you receive or extract it from the source systems before you apply any changes.

Let us compare a real lake with a data lake to make it simpler to understand for you. As you know, a real lake receives water from its various tributaries (branches or streams); it may also receive water from direct rain from the sky. Similarly, a data lake receives structured and unstructured data from various business applications, IoT devices, social media platforms, and so on. The data flowing into a data lake can be a real time stream of data or a batch of data coming in at a set frequency. Data lakes allow us to democratize the data within an organization so that whoever wants to access any data can do so by accessing it from a data lake instead of going to different source systems directly, each of which may have a different security or access model.

A data lake is also a very cost-effective way to efficiently store a large volume of data that can easily be accessed and maintained. Cloud-based data lakes have emerged as the preferred way of hosting data lakes, as the cloud provides a lot of additional benefits, like common security and access models, unlimited data storage capacity, scalability, reliability, availability, and so forth. While a data lake brings many benefits to the table, it has its own limitations. For example, a data lake does not support transactions. This means that you cannot run insert, update, and delete queries so easily on the data lake. It also lacks data consistency. As a result, new systems have begun to evolve to satisfy the market needs; they also can do away with many of the limitations of a data lake. Delta Lake is a step in that direction. So now, let us discuss Delta Lake.

What Is Delta Lake?

Delta Lake is an open source storage layer on top of an existing data lake and brings ACID (Atomicity, Consistency, Isolation, and Durability) transactions to the data lake. Without the Delta Lake layer in a data lake, it is very difficult to maintain data integrity manually, which leaves a lot of scope for human-induced errors. Delta Lake gives integrity to the data stored in a data lake, which allows the writing of data with concurrent reads out-of-the-box without any manual intervention.

Scalable metadata is another feature of Delta Lake and allows us to handle a large volume of metadata without any issues. This is a very necessary feature when handling a Big Data workload, as it may have a large volume of data that has a large volume of metadata as well. Delta Lake supports the scaling of metadata by using Spark's distributed computing power. Delta Lake stores data in Apache Parquet format, which is an open source format. It inherently supports efficient data compression and encoding schemes.

Delta Lake also provides versioning of data. This makes a very good feature called time travel possible. With it you can go back into the history and look at the various versions of your data. It gives a view of all the changes that have happened to your data since its ingestion into Delta Lake. You can also decide to undo changes that are not as per your requirements and restore any previous version of your data. Delta Lake also supports merge, update, and delete operations on data. This brings a lot of efficiency and speed to data-related operations in Delta Lake.

What Is Apache Spark?

Apache Spark is an open source distributed computing framework. It is a unified analytics engine for large-scale data processing. It can run your workload at a faster rate than Hadoop as it does all the computing in memory. It also provides multiple APIs to build the data applications using it. It supports Java, Scala, Python, R, and SQL as programming languages. It can also easily process both batch data and streaming data. It has a separate library for graph processing called GraphX. Similarly, it provides MLlib for machine learning–based workloads. Another benefit of using Apache Spark is that it can run everywhere, which includes Hadoop, Apache Mesos, Kubernetes, Standalone, or in the Cloud. It can also access various data sources for building data-driven applications.

Apache Spark has become the de facto Big Data analytics tool. Since it can read many different file formats, it can be used to process structured, semi-structured, and unstructured data from data lakes. Apart from data lakes, it can connect to many other data sources with ease. As it does in-memory computation, it is faster than Hadoop-based systems, which use MapReduce-based jobs to carry out large-scale data processing. Both the Azure Synapse Analytics and Unified Data Analytics platforms from Databricks use Apache Spark as the compute engine. For Azure Synapse Analytics,

Microsoft has done its own implementation of Apache Spark, and they call it Synapse Spark. They have also introduced free and open source .NET APIs for Apache Spark to bring a very large pool of .NET resources into the data landscape. Like that, Databricks also uses Apache Spark as the foundational technology on top of which they have built their Unified Data Analytics platform. So, Spark has become a must-have tool for data lakehouse implementations.

So far, we have discussed data warehouse, data lake, Delta Lake, and Apache Spark at a very high level. Now let us discuss what a data lakehouse is, as we have already discussed the major ingredients required for a data lakehouse.

What Is a Data Lakehouse?

A data lakehouse is a relatively new concept and a big paradigm shift from the old thought process around building a separate data lake and data warehouse. A data lakehouse is an amalgamation of the best components from both data lakes and data warehouses. A data lakehouse implements data structure and data management features from data warehouses into a cost-effective storage like a data lake. It tries to combine the best from both worlds—data lake–based Big Data analytics and a data warehouse. This is an emerging field, and there are continuous innovations happening, which results in frequent updates to the concepts, terminologies, tools, and technologies. The data landscape has evolved from data warehouse to data lake to data lakehouse over a long period. It is predicted that during the next decade, the data lakehouse will become one of the most widely used systems in the data landscape field due to the benefits it will bring to businesses. It started with data warehouses in the late 1980s and evolved into data lakes in 2011 due to the increasing need to accommodate structured, semi-structured, and unstructured data. It took around 30 years to evolve from data warehouse to data lake, but the data lake to data lakehouse evolution happened at a faster pace, occurring by 2014 (Figure 2-1).

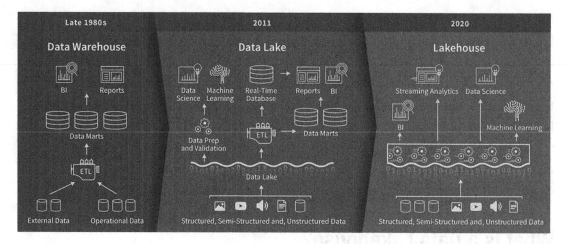

Figure 2-1. *Evolution from data warehouse to data lake to data lakehouse.* (*Source:* https://databricks.com/blog/2020/01/30/what-is-a-data-lakehouse.html)

Characteristics of a Data Lakehouse

With the preceding introduction to data lakehouses under our belts, let us discuss some of their important characteristics. These characteristics will help you to understand the various features and concepts of data lakehouses in a more detailed manner.

Various Data Types

Data lakehouses support various data types, including structured data, semi-structured data, and, most important, unstructured data. This feature is derived from data lakes, as data lakehouses use data lakes to store various types of data in the most cost-effective way, with the option to have unlimited scalability. Data lakehouses support the storage and analysis of text, images, audio, video, etc., which are generally the data types used in any modern application. The ability to derive insights from unstructured data is the most important aspect of data lakehouse implementation.

AI

To derive meaningful and useful insights from unstructured data like text, images, audio, video, etc., you will have to use various AI tools and technologies. Data science, machine learning, deep learning, **n**atural **l**anguage **p**rocessing (**NLP**), and so forth are some of the important AI workloads that are supported by data lakehouses. You can use multiple tools to run these workloads on your data, and all these tools can be used against a single data repository—a data lakehouse. Support for various types of AI workload is an extremely important feature of data lakehouses.

Decoupled Compute and Storage Resources

As we saw earlier in this chapter while discussing modern data warehouses, we know that the decoupling of compute and storage resources provides many benefits. These benefits are brought forward to data lakehouses as well. You can scale your storage and compute resources independent of each other. Since both storage and compute resources use independent clusters, you can scale them to support many more concurrent users with Big Data workloads. This also means that whenever you are not using your compute resources, you can bring down those clusters to save money, and your data will remain intact in your storage resources.

Open Source Storage Format

A data lakehouse can store your data in multiple storage formats. Generally, Apache Parquet, which is an open source file storage format, is widely used. There are many tools and libraries that can be used to directly access the data stored in Parquet format. You can use appropriate APIs as per your comfort level from Scala, R, Python, and so on. These are standardized APIs that provide multiple ways to access the data efficiently.

Data Analytics and BI Tools

There are many data analytics and BI tools that can directly access the source data from a data lakehouse. That means that there is no need to maintain two copies of data—one each in the data lake and data warehouse. That data analytics and BI tools can directly access the source data increases the freshness of the data, which allows businesses to make data-driven decisions in a timely manner and which may result in more efficient and faster use of available data within an organization.

ACID Properties

We discussed Atomicity, Consistency, Integrity and Durability (ACID) properties and their importance in Chapter 1. This feature is derived from Delta Lake and plays a crucial role in data lakehouse implementation. A data lakehouse will have multiple users reading the data through various data analytics and BI tools. In addition, there will be many data pipelines that will be reading and writing data in parallel. This is possible due to the ACID properties of transactions, which ensure data consistency and data integrity for each read or write transaction being executed against the data lakehouse.

Differences Between a Data Warehouse and a Data Lakehouse

As we have discussed previously, the data lakehouse has evolved into its current form from the data warehouse. There are some differences between a data warehouse and a data lakehouse that are important to know to properly understand this evolution process.

Architecture

The modern data warehouse includes many different components to provide end-to-end solutions for data analytics workloads. You need a data ingestion tool or an ETL/ELT tool to ingest the data from disparate source systems. The ingested data will be stored in data lake storage in multiple zones based on its processing status. You will also need a separate data processing engine like Databricks to aggregate and summarize the data, which will then be pushed into a relational database, which acts as a warehouse for the data. From there, the data will be exposed to consuming applications and various reporting, BI, data analytics, and data visualization tools. This is a typical modern data warehouse architecture consisting of multiple components.

A data lakehouse differs in many architectural aspects. Instead of having different tools for various components, it tries to combine all those components into a single system. For example, Azure Synapse Analytics includes Synapse Pipelines, Synapse Spark, Azure Data Lake Storage, Delta Lake, Synapse SQL, and Integrated Power BI workspace along with support for Azure Machine Learning in a single system. A data lakehouse will have many architectural components in a single system to provide the best of both worlds—data warehouse and Big Data analytics. A data lakehouse does not require copying the data from the data lake to a relational database, which is required for a modern data warehouse. As the data stored in embedded data storage can be queried directly using SQL engine or Spark engine, you achieve faster insights into your data.

Access to Raw Data

A data lakehouse allows you to store your data in various zones in your data lake storage. For example, you can store raw data in the Landing Zone in its original format. After applying business transformations, you will be able to store the curated data in the Curated Zone. A data warehouse does not give this option. So, data teams will not have direct access to source data in its original format from a single store. Access to raw data in its original format is very important for data analysts, data scientists, and data engineers. It allows them to explore the raw data to make appropriate decisions for their respective work areas. There is high potential that a data scientist will unearth new patterns from the raw data rather than the curated data, as the curation process will make some changes to the original data. So, this is an important differentiating factor for a data lakehouse. A data warehouse will have partial data stored in it that is required by data analytics workloads. Compared to that, a data lakehouse will have all the data stored in it in one place.

Open Source vs. Proprietary

Another big difference to note between data warehouses and data lakehouses is that most of the data warehouse products are proprietary while data lakehouses support open standards. That means that data lakehouses support open source tools and technologies supported by a large community of contributors. Delta Lake, Apache Parquet, Apache Spark, and more which are open source systems and are used in data lakehouses. This means that it will be cost effective to use those tools for a data lakehouse, and it keeps the option of using extended tools based on your specific requirements pretty much open for you. However, proprietary data warehouses cannot give that option to you.

Workloads

A data warehouse will support traditional workloads like reporting, BI, data analytics, data visualization, and so forth. The data lakehouse supports all the mentioned workloads, but additionally it can support AI-related workloads in the area of data science, including machine learning, deep learning, natural language processing, and so on. These workloads are supported to some extent in the modern data warehouse, but a data lakehouse can provide tighter integration with all these modern AI workloads.

Query Engines

With a data warehouse, you will have to rely on SQL-based query engines. Even though modern data warehouses can support other query engine types, in order to query the data stored inside a relational database in a data warehouse, an SQL-based query engine is required.

Thus, the data warehouse sort of locks you in as far as the query engine is concerned, as you will have to use the proprietary query engine supported by the data warehouse you have purchased. With a data lakehouse, you will have many options beyond the SQL-based query engine with which to query the data. Apache Spark is a very fast and efficient tool that is used in data lakehouses as the query engine. There is no query engine lock-in for data lakehouses as Apache Spark is an open source tool.

Data Processing

Data processing in a data warehouse is not cost effective as compared to doing so in a data lakehouse. Data processing in a data warehouse uses the Massively Parallel Processing (MPP) architecture, which is costly compared to open source tools like Apache Spark, which is used in data lakehouses for the same purpose. MPP architecture is a node based architecture with a control node and one or more compute nodes which are responsible for running your queries in parallel. We will discuss MPP in subsequent chapters in more detail. A data lakehouse takes advantage of the benefits of distributed computing architecture. The cluster required to run Spark-based data processing workloads can be configured using commodity hardware as well.

Real-Time Data

The data warehouse stores historical data, while the data lakehouse can store historical data as well as real-time data. This is possible because the data lakehouse supports data-streaming options out of the box. Due to the increase in the use of various IoT devices and social media, support for real-time data streaming has become very crucial for any data-related projects. A data lakehouse allows you to stream the data directly into its embedded data lake storage; this is not directly nor easily possible for a data warehouse to support.

There are many other differences between a data lakehouse and a data warehouse that we have not yet discussed. Many data warehouse products are still not available in the cloud. Even though modern data warehouses are primarily cloud-based products, many traditional data warehouse products require physical infrastructure and are not offered as Platform-as-a-Service (PaaS). Data lakehouses are inherently cloud based and hence provide more flexibility and efficiency. Cost effectiveness is another difference between the two. A data lakehouse is more cost effective compared a data warehouse. A data warehouse is very old technology that has evolved with time, and so there are many reliable and robust products in the market. The data lakehouse is a relatively new and quickly evolving concept in the IT arena. There are some features that are not supported by data lakehouses, but it is fair to assume that the future belongs to data lakehouses.

Examples of Data Lakehouses

Now let us see what products or platforms are available in the market for data lakehouses. As it is a relatively new paradigm, there are not many offerings. All big public cloud companies are trying to pitch at least one product or platform in the data lakehouse space. Each of these products or platforms has its own pros and cons. Since the technology in this area is changing and evolving at a rapid pace, cloud companies are also trying to be at the forefront of the innovation in this niche. Google Big Query and AWS Redshift Spectrum have some of the features required for a data lakehouse. However, Azure Synapse Analytics and Databricks' Unified Data Analytics Platform have the majority of the features required in a true data lakehouse. So, let us look at these in more detail. We will start with Azure Synapse Analytics as that the main topic of this entire book. We will follow it up with a high-level view of Databricks' Unified Data Analytics Platform.

Azure Synapse Analytics

Microsoft believes that enterprise analytics must work at a massive scale on any kind of data, be it raw, refined, or curated. This requires that you build Big Data and data warehouse technologies together in a single system, which is like a data lakehouse. So, in that sense, Azure has a very strong platform to offer in the space of data lakehouses. Even though Microsoft does not mention it to be a data lakehouse in its official documentation, it does come with all the features of a data lakehouse.

Basically, Microsoft has combined multiple Azure data services into a single platform: Azure Synapse Analytics. For example, Synapse Pipelines are derived from the existing Azure Data Factory service, which allows you to integrate various data sources and can also orchestrate various data pipelines. Similarly, Microsoft has provided seamless integration with Power BI Workspace within Azure Synapse Studio. As you know, we need a data lake component for a data lakehouse, and Microsoft has used their existing Azure data lake Storage Gen 2 as the data lake for Azure Synapse Analytics. It supports Azure blob storage as well. However, Azure Synapse Studio is a new web-based tool and central workspace that allows you to carry out many tasks or actions in Azure

Synapse Analytics through the GUI. If you have seen the Azure Data Factory GUI then mostly you will be at home when you log in to Azure Synapse Studio for the very first time, as the look and feel as well as the navigation options are almost similar and logical.

In November 2019, during the Microsoft Ignite event, Microsoft announced the launch of the Azure Synapse Analytics platform. It is considered a robust analytics platform with limitless scale. Microsoft has already rebranded their Azure SQL data warehouse service as Azure Synapse Analytics. But please note that Azure Synapse Analytics is just not a rebranded Azure SQL data warehouse; it is definitely more than that, and most of the new features added to it—like Azure Synapse Spark, Synapse SQL including Serverless SQL Pool and Dedicated SQL Pool, Azure Synapse Pipelines, Azure Synapse Studio, and Azure Synapse Link.

Databricks

Databricks is really a data and AI company, as they say themselves. This company has made a huge name for itself in a very short period since its inception in 2013. The company is founded by the inventors of Apache Spark, Delta Lake, and MLFlow. Databricks' Unified Analytics Platform is considered a pioneer in the area of data lakehouses. Databricks brings together data science, data engineering, and data analytics into a single unified data platform. Thus, Databricks enables all data teams to work together and collaborate closely, which results in faster innovation. They continuously evolve the platform with innovative services and features. Delta Lake, which is a storage layer on top of any data lake implementation, gives the ability to insert, update, and merge data without impacting data consistency and data integrity. Delta Lake was also originally developed by Databricks, and later they open-sourced it through the Linux Foundation. MLFlow is an open source platform developed by Databricks to manage the complete end-to-end machine learning life cycle with enterprise-level scalability, security, and reliability. They offered a Managed MLFlow as well, which is built on top of MLFlow.

Databricks' Unified Data Analytics platform has many features of a data lakehouse (Figure 2-2). Their data processing engine, which is built on top of Apache Spark, is available on both Microsoft Azure and Amazon Web Services. Their platform relies on a cloud-based data lakes being provided by respective cloud service providers. Their Databricks platform provides Delta Lake on top of the existing cloud data lake. It stores data in Apache Parquet format and supports ACID properties for data integrity and data consistency. Apart from this, recently they have introduced SQL Analytics as a new product, which allows its customers to operate a multi-cloud data lakehouse that gives data warehouse–like performance at data lake economics. It also integrates with various BI tools like Microsoft Power BI, Tableau, Qlik, and more. All these features make their platform a perfect fit as a data lakehouse.

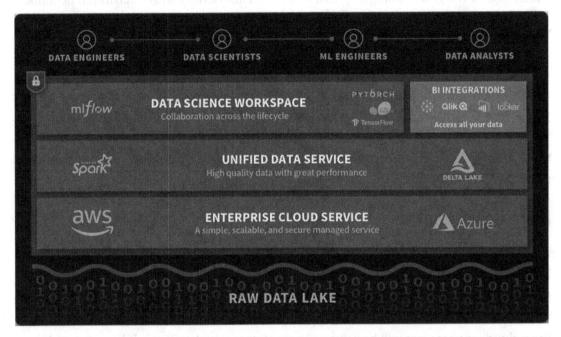

Figure 2-2. *Databricks' Unified Data Analytics Platform. (Source: https:// databricks.com/product/unified-data-analytics-platform)*

Benefits of Data Lakehouse

There are many benefits of using a data lakehouse as compared to having just a data lake or a data warehouse or having both but implemented as separate systems. Since data lakehouses pick up the best architectural components and features from both worlds, they have an upper hand in terms of features on offer in the most cost-effective way. Let us try to understand some of the important benefits of using data lakehouses.

Support for All Types of Data

The data lakehouse supports all types of data, including structured data, semi-structured data, and unstructured data. Having the ability to generate insights and intelligence from text, audio, video, images, and so on is one of the most important benefits you can take out of implementing a data lakehouse. Even though this can be achieved via data lake implementation, a data lakehouse comes with many other advantages over a data lake. In order to get the full benefits out of your data, you will have to implement a separate data warehouse system, which is not the case with the data lakehouse, as both data lake and data warehouse converge into a single system so the data lakehouse can get maximum benefits.

Time to Market

A data lakehouse will improve the time it takes to get various products and offerings to market. You will be able to go to market very fast if you are using a data lakehouse, as you will have a single system with which to generate insights and intelligence, which will speed up your data-driven decision-making process. You do not have to rely on more than one system to generate insights and intelligence. This can give a competitive advantage to you as well.

More Cost Effective

As you are going to implement and maintain only one system that will have the best of both worlds—data lake and data warehouse—you will have some cost benefits as well. Costs associated with the implementation of the system as well as ongoing maintenance will get reduced as many data-related processes will be more streamlined and some of the redundant processes will be removed completely, which will result into cost savings for you. Your data storage cost should also get reduced as a data lakehouse uses relatively cheap blob storage like AWS S3, Azure data lake storage, and others.

AI

Machine learning—and particularly deep learning—has many use cases in which you can very easily generate insights and intelligence out of unstructured data. For example, text, audio, video, images, and so forth have many hidden insights and patterns that can help businesses to make appropriate decisions quickly if they are able to analyze and make sense out of unstructured data. That is exactly how a data lakehouse can give immense benefits. Data analysts, data engineers, and data scientists like to have access to raw data to make more efficient use of the available data. Even though it is unstructured data, a data lakehouse can provide robust governance, single security model or access method, versioning, ACID properties, etc. that are crucial for any AI project, as it is well known that you need a large volume of data for an AI model to give you appropriate results.

Versioning is an extremely important aspect for unstructured data being processed by AI models. Data scientists will have to check multiple models before selecting a specific model for production implementation. Even a single model should be tested with multiple sets of data. Versioning of data will help to reproduce the same output for an AI model to prove its accuracy and precision. The data lakehouse supports versioning for unstructured data, which are primary inputs for many of the AI models. So, ideally, your data science team will be more than happy to have a data lakehouse instead of two different systems, as it will make them more efficient and productive.

Reduction in ETL/ELT Jobs

If you have two different systems, you will have to run ETL/ELT jobs to load data in both of the systems. However, with a data lakehouse that is not required. You will load into the data lakehouse once, and then you will have the option to run your queries directly against the ingested data. There is no need to load the data to the data warehouse from the data lake, which is the case if you have two different systems in your data platform. This will result in a reduction in ETL/ELT jobs, and it will make the remaining ETL/ELT jobs more streamlined and efficient.

Usage of Open Source Tools and Technologies

Many of the tools and technologies used in a data lakehouse are open source. That means that there is no license cost associated with their usage. Most of these open source tools and technologies are supported by a very large group of active contributors who continuously update and upgrade those tools and technologies. That way, you will get the benefit of faster and quicker updates and upgrades to your data lakehouse for those components. For example, it uses an open source storage format called Apache Parquet for storing data in blob storage. Additionally, it uses Delta Lake, which is an open source storage layer on top of the existing blob storage and provides features like ACID properties and support for transactions for data stored in the data lakehouse.

Efficient and Easy Data Governance

Data governance will become more efficient and easier with data lakehouses. Instead of governing multiple systems, you must govern just one data lakehouse system. In a multi-system approach, you will have to track, control, and govern the usage of multiple systems and may have to define different data governance frameworks or rules based on different requirements, as each system may have its own security model with different authentication and authorization models. A data lakehouse gives you a single point of control with a common security model as well as authentication and authorization methods. This will take away many data governance woes.

Drawbacks of Data Lakehouse

There is no architecture that is perfect and has no drawbacks. So, after seeing some of the major benefits of the data lakehouse, let us explore and try to understand some of the drawbacks of data lakehouses.

Monolithic Architecture

A monolithic architecture is one in which different aspects of a system are interwoven into a one single system so tightly that it makes it very difficult to make any changes to it. Monolithic architecture will make the administration and management aspects of the system very easy, but changes are hard to implement, and it takes time and money to implement them. It also makes the system very complex. The data lakehouse is considered a monolith system as it is an amalgamation of multiple technologies, including data warehouse, data lake, Delta Lake, Spark, and so on. As the technology evolves, we will have to see how good or bad it turns out to be in the future.

Technical Infancy

The data lakehouse is in its technical infancy. It is definitely a very new concept and big paradigm shift from the old thought process. The proposed new architecture for the data lakehouse is evolving fast, and so far it has progressed well, but it is nowhere near to a mature technology like the modern data warehouse. It is just beginning, and we hope that the technology will mature at a faster pace in the future. There are some valid reasons to believe this. There are many existing data warehouse implementations being managed by seasoned data warehouse professionals, and they will resist the change until they get the majority of the technical functionalities in the data lakehouse. This resistance will make the companies offering a data lakehouse platform to innovate and deliver new features at a faster pace.

Migration Cost

Those organizations that have data warehouses already implemented will have to think of migrating to data lakehouses. Any migration will have an associated cost. We know that a data lakehouse can bring a lot of new features and speed up the data-driven decision-making processes, but before deciding to implement a data lakehouse, organizations will have to carry out a cost-benefit analysis. Organizations will have to decide if benefits offered by the data lakehouse are worth the cost involved in migrating to it. Real technical ROIs are difficult to calculate in a short timeframe. So, we will have to wait a little longer to have some clarity in this area.

Lack of Many Products/Options

There are not many products or offerings in the market for data lakehouses. Apart from Azure Synapse Analytics from Microsoft and Unified Data Analytics platform from Databricks, there are no other products that have the majority of the features of a data lakehouse. Snowflake, AWS, Google, and so forth have some sort of products in the market, but those are not mature enough to be called a data lakehouse. Compared to that, we have a lot of products on the market that can be categorized as a modern data warehouse. We can hope that going forward there will be many mature products in the market in this space.

Scarcity of Skilled Technical Resources

This is true for any new technology or widely used technology with a steep learning curve. In particular, cloud-based technologies can become popular very fast, and many organizations strive to use cutting-edge technology to stay ahead of their competitors. However, technical resources will not be so readily available in the market when the specific technology becomes very hot. All data-related job roles are in demand, which includes data analysts, data engineers, data scientists, and data architects. A data lakehouse implementation will need skilled technical resources for all these roles. It is difficult to find skilled technical resources in technologies related to data lakehouses. As the technology matures and companies continue to push their products in this area, it will attract new talent and increase the talent pool in the future.

Summary

In this chapter, we have built our knowledge on top of the discussion covered in Chapter 1. We started with data warehouses and discussed what a data warehouse is and why an organization needs a data warehouse. Over a period, the traditional data warehouse evolved into the modern data warehouse; cloud technologies have played a big role in this evolution. There are some important differences between a traditional data warehouse and a modern data warehouse, and we covered those differences briefly. The data landscape has been evolving continuously, which is evident in the evolution from data warehouse to data lake to support unstructured data like text, images, audio, video, and more. This was an important milestone in the evolution process. The emergence of Big Data has resulted in many systems existing in the Big Data analytics space. Data lakes have played a pivotal role in that.

The evolution process from data lake to data lakehouse was faster compared to the evolution process from data warehouse to data lake. To understand data lakehouses and their related concepts, it is necessary to understand what a data lake is and what a Delta Lake is. We discussed these foundational concepts before examining what a data lakehouse is and covering important characteristics of data lakehouses. A data lakehouse provides many benefits over a data warehouse or a data lake, and we discussed some of those benefits. There are many early adopters of the data lakehouse paradigm, and Azure Synapse Analytics and Databricks' Unified Data Analytics platforms are the best examples of a data lakehouse commercially available today. No technology is perfect, and hence we covered some of the drawbacks of data lakehouses as well.

With this chapter, we have built foundational knowledge to get into a more detailed discussion about Azure Synapse Analytics. The remaining chapters of this book will fully concentrate on Azure Synapse Analytics. In the next chapter, will give a detailed introduction to Azure Synapse Analytics. We will look at how Azure Synapse Analytics is different from Azure SQL Datawarehouse and why we should learn it. We are also going to cover some of its key features and key services capabilities.

See you in the next chapter, in which we will be beginning our journey toward learning Azure Synapse Analytics!

CHAPTER 3

Introduction to Azure Synapse Analytics

In Chapters 1 and 2, we discussed some of the important concepts around core data, various types of data, their characteristics, relational and non-relational data, and so forth. We also covered basic and conceptual knowledge regarding traditional data warehouses, modern data warehouses, and data lakehouses. Based on that foundation, we are now ready to take our first step in our journey toward learning Azure Synapse Analytics, which is the main topic of this book.

We got a high-level introduction to Azure Synapse Analytics as part of Chapter 2 while discussing examples of data lakehouses. However, as part of this chapter, we are going to get a more detailed and deeper introduction to it. Azure Synapse Analytics is an analytics platform, as its name suggests, but it can also be used as a data lakehouse as well as a modern data warehouse because it contains multiple technical ingredients that meet the criteria of those different systems independently. Azure Synapse Analytics evolved into its current state over a long period.

By now we know that Azure Synapse Analytics evolved into its current form by converging the data lake and modern data warehouse concepts. Historically, relational databases were used to store transactional data as well as analytical data.

What Is Azure Synapse Analytics?

Let us look at how Microsoft defines Azure Synapse Analytics in its official documentation online:

© Bhadresh Shiyal 2021
B. Shiyal, *Beginning Azure Synapse Analytics*, https://doi.org/10.1007/978-1-4842-7061-5_3

"Azure Synapse Analytics is a limitless analytics service that brings together data integration, enterprise data warehousing and big data analytics. It gives you the freedom to query data on your terms, using either serverless or dedicated resources—at scale. Azure Synapse brings these worlds together with a unified experience to ingest, explore, prepare, manage and serve data for immediate BI and machine learning needs."

In simple words, Azure Synapse Analytics is an analytics service offered by Microsoft on Azure, which is its public cloud offering in the market. It brings together the following three components:

1. Data integration

2. Enterprise data warehouse (EDW)

3. Big data analytics

Data integration is the process of ingesting data from disparate data sources and integrating them into a central repository. Azure Synapse Analytics supports this process though Synapse Pipelines, which gives no-code or low-code data ingestion options, which improves efficiency. Apart from this, it also supports streaming the data to take care of real-time data ingestion requirements. For data integration purposes, Microsoft has tightly integrated Azure Data Factory into Azure Synapse Analytics as Synapse Pipelines. With more than 90 different readily available connectors, Synapse Pipelines provides a comprehensive toolset to integrate and ingest data from multiple data sources.

Microsoft has rebranded Azure SQL Data Warehouse as Azure Synapse Analytics. That means that whatever capabilities of the enterprise data warehouse that were available in Azure SQL Data Warehouse have been carried forward to Azure Synapse Analytics. However, it must be noted that Azure Synapse Analytics is just not a rebranded Azure SQL Data Warehouse—it is a lot more than that. In addition to bringing the proven foundation of the SQL engine from Azure SQL Data Warehouse, there are many enhancements that were added to the EDW capabilities in Azure Synapse Analytics. If you are confused between Azure SQL Data Warehouse and Azure Synapse Analytics, do not worry too much, as shortly we will cover some of the key differences between the two.

Big Data analytics is the third most important component in Azure Synapse Analytics. It provides the capability to query data stored in data lakes very easily. This has been made possible because Azure Synapse Analytics supports Apache Spark as a distributed query engine. It allows you to query the structured, semi-structured, and unstructured data stored in your data lake with the same service that you use to build your EDW. Data lake exploration capability is an important characteristic of data lakehouses, which are available in Azure Synapse Analytics as well. These allow your data teams to explore and generate insights and intelligence from the data available in the data lake in a much faster way, as they will not have to wait for data to get loaded into the data warehouse.

Another important phrase to note down is the freedom to query your data on your own terms. You have two options: serverless resources and dedicated resources. Azure Synapse Analytics provides Synapse SQL On-Demand and Synapse Spark options when it comes to using serverless resources. For dedicated resources, you must use Synapse SQL Pool, which is a set of reserved resources for your SQL workloads. Based on your requirements and the type of workloads, you can decide which type of compute resources should be used. This gives a lot of flexibility to your data teams, as they will have more options to run their workloads in the most cost-efficient and fast way.

Azure Synapse Analytics also brings a unified experience for data ingestion, data exploration, data preparation, data management, and serving the data to the consuming layer in terms of BI- and machine learning–related requirements. This is made possible by Synapse Studio, through which your data team can carry out all types of data-related tasks without leaving the interface provided by Azure Synapse Analytics. We will go through all these topics in much more detail in this chapter and subsequent chapters, but first let us try to understand some of the key differences between Azure Synapse Analytics and Azure SQL Data Warehouse.

Azure Synapse Analytics vs. Azure SQL Data Warehouse

We covered some of the important differences between a data warehouse and a data lakehouse in Chapter 2. Azure Synapse Analytics can be used as both a data warehouse and a data lakehouse. But you may ask what happened to Azure SQL Data Warehouse, as it was a separate service that was available in Azure. It is a little confusing, as Microsoft has officially rebranded Azure SQL Data Warehouse as Azure Synapse Analytics. That means that Azure SQL Data Warehouse is included in Azure Synapse Analytics. All the existing customers of Azure SQL Data Warehouse were migrated to Azure Synapse Analytics. However, please note that this includes only enterprise data warehouse capabilities; for additional capabilities like data lakes, Synapse Pipelines, Synapse Spark,

Synapse Serverless SQL Pool, and so forth, customers will have to opt in and pay for them separately. Based on your requirements, you can decide either to use one or more services or to not use any of the additional services. Thus, Azure Synapse Analytics offers your erstwhile Azure SQL Data Warehouse capabilities, as those were available to you before the launch of Azure Synapse Analytics, without any price changes for existing customers, but provides additional options that you can decide to opt into based on your requirements.

There are many additional services in Azure Synapse Analytics that were not present in Azure SQL Data Warehouse. Azure SQL Data Warehouse does not exist any longer as an independent service. If you want to use Azure SQL Data Warehouse capabilities, then you must go to Azure Synapse Analytics and use a Dedicated Synapse SQL Pool, just as you were using it previously in Azure SQL Data Warehouse. Azure SQL Data Warehouse was a cloud-hosted modern data warehouse built using proprietary technologies from Microsoft. Compared to that, Azure Synapse Analytics is a comprehensive data analytics platform that brings enterprise data warehouse and Big Data analytics at a cloud scale into a single platform. Thus, Azure Synapse Analytics offers far more services and benefits compared to Azure SQL Data Warehouse. The biggest differentiator would be Synapse Spark, which is Microsoft's own implementation of an open source Apache Spark engine and was not available in Azure SQL Data Warehouse.

Why Should You Learn Azure Synapse Analytics?

If you are reading this book, it means that you are interested in learning Azure Synapse Analytics. It is next-gen, cutting-edge technology that combines the power of enterprise data warehouses and Big Data analytics in a single system. It is a perfect fit for data lakehouses, which are an emerging trend in the market that is developing at a very rapid pace due to continuous innovations happening in the data analytics field.

As discussed in Chapter 1, data is the new oil. It is driving innovations in various industry verticals across the globe. The importance of data-driven decision-making processes cannot be stressed enough. It is the need of the hour. As a result, many organizations have started to implement modern data platforms or modernize their existing legacy data platforms. This is driving strong demand in the global market for data-related professionals. Data analysts, data engineers, data scientists, data architects, BI developers, and so on are in high demand in the market.

Azure Synapse Analytics offers multiple services within it that can be used by various data professionals based on their roles within a project. Data analysts who want to analyze the data stored in a data lake can easily do so by querying the data using Synapse Studio. Additionally, Power BI Workspace is tightly integrated with Synapse Studio. It allows you to generate various reports, dashboards, KPIs, and so forth using myriad visualization options without leaving Synapse Studio. Data engineers are primarily responsible for building the data pipelines to ingest the data from disparate data sources. They can use Synapse Pipelines to build those data pipelines and ingest the data into a data lake. Data engineers are also responsible for cleansing and transforming the ingested data before it is consumed further by downstream applications or users. They can use Synapse Studio Notebooks to write their queries for cleansing the data and applying various transformations. Data scientists are required to apply various machine learning algorithms to predict future outcomes. They can also use Synapse Studio along with Azure Machine Learning to carry out activities related to a data science project.

Main Features of Azure Synapse Analytics

Now, let us explore some of the important features of Azure Synapse Analytics. This will help us to understand it in a little more detail and see what it has to offer as a data analytics platform.

Unified Data Analytics Experience

Azure Synapse Analytics provides a unified experience when it comes to building end-to-end data analytics solutions. This unified data analytics experience is provided through Synapse Studio. This is a newly introduced cloud-native web-based GUI tool in Azure Synapse Analytics. Synapse Studio provides a central workspace to carry out tasks related to data ingestion, data preparation, data exploration, data management, data warehousing, data lakes, AI, and so forth.

Data engineers can quickly build and manage data pipelines using a no-code or low-code visual environment by using Synapse Studio. The experience is similar to that of Azure Data Factory. Database administrators can write SQL queries to optimize and automate database-related tasks using Synapse Studio. Similarly, data scientists can rapidly conceptualize and build proofs of concept, as Synapse Studio also integrates Azure Machine Learning into it. Data analysts who are required to analyze data and then build reports, dashboards, KPIs, and so on using Power BI can use Synapse Studio. This is possible as you can tightly integrate Power BI Workspace into Synapse Studio directly. Data analysts will have a similar experience while using the Power BI integrated workspace.

Thus, Azure Synapse Analytics provides a unified data analytics experience to various types of roles or users within your data teams. It also enables greater collaboration among different data teams within your organization. This results in faster development and more innovations as everyone is using the same platform to perform their specific tasks. Thus, Azure Synapse Analytics in a true sense provides a unified analytics experience. There is no need to utilize various tools and have different GUI experiences when it comes to data analytics–related activities.

Powerful Data Insights

Azure Synapse Analytics provides tight integration with Power BI as well as Azure Machine Learning. These integrations expand the discovery of insights and intelligence from all organizational data. Machine learning engineers can easily build and apply machine learning models on all data since they have access to it without having to move it from one place to another for the given purpose. These integrations also reduce project development time for data analytics projects, including BI and machine learning projects. Azure Syanapse Analytics can cover all important organizational data irrespective of where the data is stored, as you can use data stores in Dynamics 365, Office 365, and other SaaS services. The only caveat here is that the source data service should have support for Open Data Initiative.

Open Data Initiative is a platform jointly developed by Microsoft, Adobe, and SAP that provides a single comprehensive view of your data. It provides artificial intelligence–generated insights from all your data to build more powerful systems for your organization. Please refer to `https://www.microsoft.com/en-us/open-data-initiative` for more information on Open Data Initiative. Thus, Azure Synapse Analytics is very useful in generating very powerful data insights, including BI, AI, and advanced data analytics, which will help any organization to make data-driven decisions.

Unlimited Scale

Azure Synapse Analytics provides an unlimited scaling option to get insight from all the data irrespective of where it is stored, whether in data warehouses or data lakes. You can easily query structured, semi-structured, and unstructured data using Azure Synapse Analytics, even though the volume of your data may be in petabyte size. This is possible as it supports limitless scaling. As we know, the data lake is the most cost-effective and highly scalable storage, and it is part of Azure Synapse Analytics. That is why Azure Synapse Analytics can easily support petabyte-size data volume, which provides unlimited storage. Similarly, it is also possible to easily scale the compute resources in Azure Synapse Analytics for Synapse SQL as well as Synapse Spark clusters.

A Serverless Synapse SQL Pool does not require you to provision any compute resources up front. Internally, Azure Synapse Analytics will ensure it provisions the necessary compute resources based on the query you submit to it. Similarly, a Dedicated Synapse SQL Pool can also be scaled by increasing the allocated Data Warehouse Units (DWUs) to it. A Synapse Spark pool can support a maximum of 200 nodes within a single cluster. Thus, be it storage or compute, Azure Synapse Analytics provides options to scale to match the requirements of your workload.

Security, Privacy, and Compliance

Cloud security and data privacy are always challenging to manage. Data breaches are becoming common. It is necessary to keep your data safe with advanced security and data privacy features to avoid data breaches as well as to meet various regulatory and legal requirements. Thus, identifying and implementing best practices for security and data privacy are some of the toughest tasks in the IT security and privacy domain.

Azure Synapse Analytics provides some advanced security features, which include automated threat detection as well as always-on encryption. It is also necessary to have fine-grained access control, and that is made available in Azure Synapse Analytics through features like column-level security as well as row-level security. Additionally, Azure Synapse Analytics supports column-level encryption and dynamic data masking. It helps in protecting sensitive information in real time. Azure Synapse Analytics provides many security and data privacy features that will allow you to define a comprehensive and long-term strategy for securing the data and meeting various compliance-related obligations around data privacy.

HTAP

Through the Azure Synapse Link feature, Azure Synapse Analytics provides a way to get instantaneous insights from your operational data. This is made possible by implementing an innovative concept called the Hybrid Transaction/Analytical Processing (HTAP) model, which is a hybrid approach that takes benefits of both worlds—OLTP and OLAP—as its name suggests. Azure Synapse Link is a simple, low-cost, cloud-native implementation of HTAP that gives you the ability to generate insights into your business data immediately. This in-the-moment insight is the need of the hour as everything is going toward real-time or near real-time. Various social media platforms and e-commerce websites have made us expect instant gratification and realization. Now, this is being seen in the data analytics platform as well.

Azure Synapse Link creates a simple bridge between your operational database services and your analytics platform. The bridge integrates both so easily that you can generate instant and real-time insights from your transactional or operational data without moving it to an analytics platform. It also does not create any additional load on your operational databases. This means that you can have a no-ETL solution for your data analytics platform without requiring any additional compute resources. At the time of writing this book, Azure Synapse Link supports Azure Cosmos DB. However, Microsoft has a clear roadmap to implement it for other Azure databases, including Azure SQL Database, Azure Database for PostgreSQL, Azure Database for MySQL, and others, in the near future.

So, we have covered some of the most important features of Azure Synapse Analytics that make it the most compelling, promising, and comprehensive unified data analytics platform. With that knowledge in hand, let us discuss some of the key service capabilities of Azure Synapse Analytics.

Key Service Capabilities of Azure Synapse Analytics

Azure Synapse Analytics has multiple service capabilities that are important to know for users. These service capabilities provide a detailed view of various technical capabilities of Azure Synapse Analytics. We are going to discuss some of the key service capabilities one by one. This will give you a detailed understanding about what is technically feasible in Azure Synapse Analytics and which service capabilities can be utilized for the given technical scenarios. So, let us get going!

Data Lake Exploration

This is one of the most important service capabilities of Azure Synapse Analytics. There are a couple of strong reasons why it is so important. First, it allows you to store relational and non-relational data together in a single place in data lake storage. Azure Synapse Analytics supports Azure Data Lake Storage (ADLS) Gen 2 as the data lake storage component. A data lake can easily store structured, semi-structured, and unstructured data in a single data lake storage. It provides unlimited storage capacity. It is the most cost-effective storage option; hence, it is very cheap to store data in data lake storage. Azure Data Lake Storage also provides various other features like advanced security, scalability, reliability, availability, and more. To explore your data easily and most efficiently, it is necessary that you store all your data, irrespective of its types and formats, in a single storage location. Once that is achieved, the next task is to check its querying capabilities.

Azure Synapse Analytics also allows you to query the data stored in the data lake easily and efficiently. This is possible through a couple of features like Synapse Studio, SQL Pool, and Spark Pool. Azure Synapse Analytics provides three different querying engines to query the data stored in data lake storage. As part of Synapse SQL Pools, it supports both Dedicated Synapse SQL Pools as well as Serverless Synapse SQL Pools for querying the data stored in data lake storage. Additionally, it is also possible to use Synapse Spark as data-querying tool.

With Synapse Studio, it is very easy to explore the data stored in a data lake, as it provides a GUI-based approach. You can navigate to the file you want to explore further. You can simply right-click on it, which brings up an option to select the top 100 rows from the selected file. It generates the T-SQL query for you automatically, which can easily be executed using a Serverless Synapse SQL Pool. There is no need to provision the compute resources required by the query engine in this case, as Microsoft will take care of this automatically based on the need of the query you want to execute. Additionally, Synapse Studio provides an option to visualize your data in a graphical manner without any additional effort on your part. This feature makes data lake exploration easier and even more interesting for the end users.

Without this service capability, you will have to define the external tables manually, which is a time-consuming process. Before defining the external tables, you will have to define database objects like external file format, credentials, and external data source. After this laborious process, you will finally be able to explore the data stored in data lake storage. Now, with Synapse Studio and its data lake exploration service capabilities, you can start your exploration exercise in just a couple of clicks. This saves time, and since it uses auto-generated scripts, it is less error-prone as well. Imagine that you have to manually explore a large file with hundreds of columns stored in data lake storage. This can easily be done with just a few clicks. These are the benefits you get while using data lake exploration service capabilities, and as a result these are considered one of the most important service capabilities of Azure Synapse Analytics.

Multiple Language Support

Azure Synapse Analytics supports multiple languages. That means that you can select your choice of language for development purposes. When you want to create a Notebook that you want to run against Synapse Spark Pool, you will have the following list of languages to choose from for your work:

1. PySpark (Python)

2. Spark (Scala)

3. .NET Spark (C#)

4. Spark SQL

Here, Python, Scala, and Spark SQL are the most common options, which you will get even if you are using Spark on any other platform, like Databricks, etc. However, .NET for Spark using C# is a new option that is not available in other platforms, but you can use it in Azure Synapse Analytics. Microsoft has open-sourced the .NET for Spark libraries, which opens opportunities for further customizations through open-source community contributors.

Apart from coding against Synapse Spark, you can also write SQL queries using T-SQL against a Dedicated Synapse SQL Pool or Serverless Synapse SQL Pool, which allows a large set of developers to use it without learning any new programing language. There is a large population of programmers who are well versed with C#.NET and T-SQL, hence it is an attempt by Microsoft to bring those people on board with Azure Synapse Analytics. Python and Scala are already two of the most popular languages that are widely used to code for Spark-based data analytics projects. Thus, by offering multiple languages for Azure Synapse Analytics, they make it easy to learn, understand, and use this unified data analytics platform.

Deeply Integrated Apache Spark

Azure Synapse Analytics is deeply integrated with Apache Spark. As we discussed in the previous chapter, Apache Spark is an open source distributed analytics platform. Microsoft has created its own implementation of Apache Spark for Azure Synapse Analytics, and that is Synapse Spark. This is the third Spark engine available in Azure after Azure Databricks and Azure HDInsight. Apache Spark is considered much faster than a traditional Hadoop-based system as it processes the data in-memory and does not require writing and reading from a storage disk, as is the case for MapReduce jobs in a Hadoop-based system. You can refer to Figure 3-1, which depicts the differences between Hadoop and Spark.

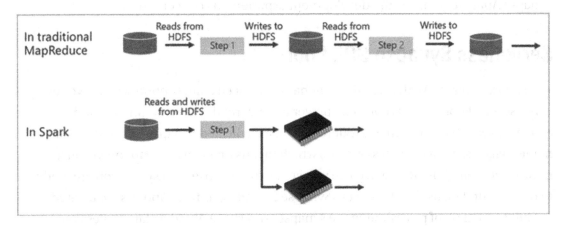

Figure 3-1. Hadoop vs. Spark. Source: *https://docs.microsoft.com/en-us/ azure/synapse-analytics/spark/apache-spark-overview. This figure comes from Microsoft Documentation, in a sub-section called "Apache Spark in Azure Synapse Analytics" dated April 15, 2020*

Azure Synapse Analytics has a built-in notebook experience that can be leveraged to develop code to be executed against Synapse Spark–based clusters. This notebook experience is built on top of the open source Nteract notebook. These notebooks are handy to build data-processing jobs and data visualizations very quickly. They allow data teams to build proofs of concept rapidly by using various Spark APIs against the data stored in data lake storage.

Azure Synapse Analytics also provides easy-to-configure options for creating, configuring, running, and managing Synapse Spark Pool, including Azure Portal, Azure PowerShell, or Synapse Analytics' .NET SDK. Synapse Spark Pool can be configured for auto-scaling, which allows you to increase or decrease the number of nodes required for your workload. This is a fully automated process; based on your workload, Spark will either increase or decrease the number of nodes it requires for data processing. There is also an option to pause the Spark Pool automatically after a preconfigured idle time. Generally, Spark Pool instances will be up and running in about two minutes for fewer than 60 nodes and in about five minutes for more than 60 nodes. Additionally, a Spark Pool instance will shut down automatically within five minutes of the last job executed if you do not keep alive at least one notebook connection. All these options will help you to save on the cost of Spark clusters.

Thus, deep integration with Apache Spark provides many benefits to the users of Azure Synapse Analytics. Spark developers are treated as first-class citizens in Azure Synapse Analytics as far as the development experience is concerned.

Serverless Synapse SQL Pool

With Azure Synapse Analytics, Microsoft has introduced a new option with Serverless Synapse SQL Pool, as far as compute power is concerned. This is an on-demand SQL pool available to you that is different from a Dedicated Synapse SQL Pool. As its name suggests, it is a serverless option, which means that you need not provision any infrastructure in order to use it. Once you create the synapse workspace, you are ready to query your data using a Serverless Synapse SQL Pool, as its endpoints are created by default at the time of provisioning the synapse workspace. Various data professional roles can use the same Serverless Synapse SQL Pool as it supports multiple use cases and can cater to many tasks performed by different data professional roles, like data engineers, data scientists, data analysts, and BI professionals.

It is important to note that you can query only data stored in your data lake or Spark tables using a Serverless Synapse SQL Pool. It is in a true sense a pay-as-you-go service, as you do not pay anything for any provisioned resources for a Serverless Synapse SQL Pool. Based on your query size, you will be charged for using a Serverless Synapse SQL Pool. It charges per terabyte of data processed by your query. It provides functionalities in familiar T-SQL syntax to query data in place without the need to copy or move the data into a specialized data store.

Now, let us try to understand some of the use cases in which you can benefit from using a Serverless Synapse SQL Pool. Data lake exploration is the most widely known use case for it. You can explore the data stored in your data lake in CSV, JSON, Parquet, and other formats by querying it to plan your data-extraction and data-processing jobs so as to generate meaningful insights. Another use case is to transform your data in a simple and efficient way using familiar T-SQL syntax so that the data can be consumed by BI, AI, and other data analytics tools. Apart from Azure Synapse Studio, there are two other client tools—Azure Data Studio and SQL Server Management Studio (SSMS)—that you can connect to a Serverless Synapse SQL Pool for your work.

Thus, the Serverless Synapse SQL Pool is an important and critical new service capability added to Azure Synapse Analytics that is useful for any data professionals working on an Azure Synapse Analytics workload. Its serverless nature and true pay-as-you-go option are the icing on the cake for its usage.

Hybrid Data Integration

Data integration is an important and strategic part of any data analytics project. Azure Synapse Analytics makes it easy to integrate various data sources and brings the data to a central storage location. It provides a facility in which to develop both ETL and ELT processes based on your data integration requirements. It is really easy to build data ingestion pipelines as it provides a code-free or low-code visual development experience. This is made possible through Synapse Pipelines, which is a core component of Azure Synapse Analytics. Synapse Pipelines are data pipelines based on an Azure Data Factory implementation within Azure Synapse Analytics that take care of the data ingestion-, integration-, and orchestration-related requirements.

Azure Data Factory is a cloud-native de facto data orchestration tool within the Azure cloud. It supports more than 95 native connectors, which will allow you to connect to various data sources very easily through the GUI. Synapse Pipelines' development experience is similar to what you get in Azure Data Factory. So, if you have already used Azure Data Factory or are familiar with it, then you will feel at home when you get into Synapse Pipelines. It is directly integrated with Synapse Studio; hence, you need not leave the Synapse interface in order to develop or manage your data pipelines, as this can be carried out from the Synapse Studio–based GUI. When you have to switch between your data pipeline and data processing code while troubleshooting any issues in either, you can easily do this from the single interface of Synapse Studio.

Synapse Pipelines also support authoring and executing data flows, which are also available in Azure Data Factory. Data flows can be designed visually to apply various data transformations on your data without writing any code. Additionally, this can be done from the Synapse workspace itself without leaving Synapse Studio. Once the data flows are authored and published, they can be executed as activities from Synapse Pipelines. These data flows will be executed in scaled-out Synapse Spark clusters. Thus, hybrid data integration through ETL/ELT as well as data flows provide a tightly coupled data integration experience in Azure Synapse Analytics.

Power BI Integration

Power BI developers are responsible for developing solutions that provide actionable insights into the data. Power BI requires various data sources to provide data to it, from which it can generate data visualizations using dashboards, reports, and so on. Azure Synapse Analytics is supported as one of the many data sources to which Power BI can connect and consume data from for various BI purposes. Azure Synapse Analytics can work as a single source of truth for Power BI reports.

Azure Synapse Analytics supports DirectQuery mode for integrating with Power BI. However, compared to other data sources that support DirectQuery mode with Power BI, Azure Synapse Analytics adds performance optimizations in terms of materialized views as well as result-set caching. Due to these optimizations, Power BI DirectQuery mode with Azure Synapse Analytics can support large source datasets along with thousands of concurrent users for Power BI reports.

Power BI App Workspaces can be directly integrated with Azure Synapse Studio. Once the necessary permission is in place, you can connect to a specific Power BI App Workspace from the Synapse Studio interface. Once you integrate the Power BI App Workspace, you will be able to carry out all the necessary tasks of developing visualizations, dashboards, reports, and so on. As part of this integration process, you will create a linked service for Power BI. Once the Power BI App Workspace–specific linked service is created, you can access that specific Power BI App Workspace from within the Synapse Studio interface.

Azure Synapse Analytics' integration with Power BI also provides for closer team collaboration. As all teammates playing different data roles on your data team can use the same Synapse Studio interface to carry out their respective tasks, it provides opportunities for closer collaboration within the data team. Be it data engineer or data scientist or data analyst, all will be using the same Synapse Studio. The data engineering team that is building a data pipeline for your project can collaborate with your team's Power BI developers to iron out any adverse impact on your downstream consuming applications. This increased team collaboration will result in faster time to market and rapid development cycles, which are ultimately beneficial for business.

Thus, Power BI's integration with Azure Synapse Analytics brings more value for both sides. The data consumption layer is the culmination of all your data efforts, as that is where your data is made really useful and meaningful for your business. Since Power BI is a service from Microsoft, there is nothing wrong in expecting a higher level of tighter and deeper integration with Azure Synapse Analytics going forward. It is also necessary to note that apart from Power BI, Azure Synapse Analytics supports integration with some of the other very well-known BI tools, which include Tableau, MicroStrategy, Cognos, Qlik, and more.

AI Integration

AI has become part and parcel of each IT offering nowadays due to its increasing importance in solving many real-life use cases. Azure Synapse Analytics provides AI integration as it provides various machine learning capabilities.

Generally, machine learning projects require you to acquire and understand the data. For data acquisition, you can use Synapse Pipelines from Azure Synapse Analytics, which is nothing but Azure Data Factory integration. Synapse Pipelines can help you to build those data pipelines that are required for your machine learning projects. Once the data is acquired and stored in persistent storage, you need to be able to understand the data, and for that you will have to use various exploratory data analysis (EDA) techniques. Here, Synapse Spark as well as a Serverless Synapse SQL Pool can help you to carry out those tasks.

Machine learning also requires you to train the selected model on the training data. Synapse Spark provides MLlib, which gives a scalable machine learning algorithm. If there is a need to use a third-party library to train your machine learning model, then that is also supported in Synapse Spark, which allows you to install and use those libraries. It is also possible to pick up a model trained on any other platform and then convert it into an internal representation supported by Azure Synapse Analytics. This saves time as it avoids having to start the process from scratch. If you decide to train your model using automated machine learning, then that too is supported in Azure Synapse Analytics. You can use Azure Machine Learning's Automated ML feature to train a set of machine learning models and select the most appropriate ML model out of it automatically.

Once a model is trained, it is deployed and scored. Azure Synapse Analytics supports batch scoring of models irrespective of where it is trained—using Azure Synapse Analytics or outside Azure Synapse Analytics. With Synapse SQL Pool, you can use the T-SQL PREDICT function to run your predictions at the same place where your data is stored. Alternatively, you can use Synapse Spark Pools for this purpose based on which ML libraries you have used for training your model. You can use notebook code to score your model in batch mode.

As part of AI integration in Azure Synapse Analytics, it is just not Azure Machine Learning that is integrated directly, as Azure Cognitive Services is also integrated with it. You can easily utilize already trained models from various cognitive services based on your requirements. Synapse Studio allows you to use cognitive services to enrich your data stored in Spark tables through the GUI. You can right-click on any Spark table in Synapse Studio and it will give a machine learning menu with two different sub-menus under it. One is to enrich your data using existing model, and that is where it allows you to select whichever Azure Cognitive Services you want to use for enriching your data.

Thus, Azure Synapse Analytics provides a tight integration with AI capabilities. Azure Machine Learning as well as Azure Cognitive Services both can be used as and when you want to based on your machine learning projects' requirements.

Enterprise Data Warehousing

Enterprise data warehousing is the core and foundational service capability of Azure Synapse Analytics. In the previous chapter, we discussed the data warehouse and its characteristics, along with certain other important topics. We also discussed how traditional data warehouses evolved into the modern data warehouse, and we even compared it with data lakehouses. Azure Synapse Analytics is originally based on Azure SQL Data Warehouse; hence, it is very obvious that all the features of Azure SQL Data Warehouse that were on offer as an Azure service are being carried forward to Azure Synapse Analytics, but with a lot of improvements and game-changing features.

You can rely on Azure Synapse Analytics to build your mission-critical data warehouse on a solid foundation. Azure SQL Data Warehouse is one of the most widely used and comprehensive modern data warehouse systems. It works on industry-recognized and proven massively parallel processing (MPP) architecture. Microsoft has continued the compute engine, which was available in Azure SQL Data Warehouse, as a Dedicated Synapse SQL Pool in Azure Synapse Analytics. It also carried forward the

Azure SQL Data Warehouse Gen2 compute data warehouse unit (cDWU). It allows you to increase or decrease the DWUs based on your workload requirements. It continues to support service-level objectives (SLO), which determine your scalability settings. SLO decides the cost and performance level of your enterprise data warehouse.

Seamless Streaming Analytics

Since its launch, Azure Synapse Analytics has enabled support for direct streaming ingestion. It provides capabilities to execute analytical queries over data being streamed to Azure Synapse Analytics. You can integrate with Azure Event Hubs and Azure IoT Hubs. Here, Azure Event Hubs can also be used for the Kafka protocol, which is one of the most widely used streaming technologies.

After its initial launch in November 2019, Microsoft made certain improvements to support seamless streaming analytics on Azure Synapse Analytics. However, the initial integration for streaming was a bit cumbersome to set up, as you had to provision Azure Stream Analytics and Azure Databricks in order to use streaming analytics. However, as per the latest changes, streaming capabilities have become seamless to onboard and manage on Azure Synapse Analytics. Now, it supports COPY Command behind the scenes for high-throughput data ingestion—and that is without having any CONTROL permission on the databases to set up the streaming jobs.

Additionally, setting up and managing streaming jobs can now be done directly from Azure Synapse Analytics without visiting different services with different user experiences. It now also supports automatic schema detection as well as automatic table creation for streaming input sources. Thus, Azure Synapse Analytics natively supports seamless streaming analytics workloads, and it has become easier to on-board and manage streaming jobs on the platform.

Workload Management

Workload management is key for any data analytics systems in which there are always competing demands for available resources. These competing demands include the following workloads:

1. Data ingestion process

2. Data processing/transformation

3. Data querying, analysis, and reporting

4. Data management

5. Exporting data

Data architects try to find a way to separate all these workloads and devise a balanced approach to manage them based on their importance and requirements. Data architects are responsible for creating a workload management scheme that can meet the following objectives at a high level:

1. Manage the available resources in the most efficient manner possible.

2. Ensure highly efficient resource utilization.

3. Maximize Return on Investment (ROI).

In the past, resource classes were used to assign memory to a query based on role membership. This means that it would depend on the user's assigned role. Once you assigned a user to a role, there was no control over the resource class and a single user could end up consuming all the memory. To avoid this situation, Azure Synapse Analytics uses the following high-level concepts for workload management for Dedicated Synapse SQL Pools:

1. Workload classification

2. Workload importance

3. Workload isolation

As part of the workload classification concept, a request is assigned to a workload group along with its importance level. Workload classification provides rich capabilities as it has various options to classify requests like label, session, and time. This is an improvement over having only a user membership–based or role-based option to classify the requests. Workload importance decides the order in which a request will get access to resources. This concept allows you to decide priorities for various competing workload demands. Workload isolation gives you the ability to keep the necessary resources reserved exclusively for a workload group to ensure execution.

Thus, workload management is one of the most important service capabilities because it allows you to manage competing workloads in the most efficient way to ensure the smooth execution of your workload so as to maximize the ROI for your Azure Synapse Analytics platform.

Advanced Security

Azure Synapse Analytics provides state-of-the-art and advanced security capabilities. It follows the same layered-defense, in-depth security strategy as do Azure SQL Database and Azure SQL Managed Instance.

The outermost layer is network security, and that includes IP firewall rules and virtual network firewall rules. These rules allow you restrict access to Azure Synapse Analytics, except for the IP addresses included in the rules. The next layer of the defense-in-depth security strategy is access management. This is controlled through various authentication and authorization options. It also allows you to define row-level security as well as column-level security for the data stored in the database. The next layer is threat protection, which is supported through SQL auditing in Azure Monitor logs and event hubs. It provides advanced threat protection (ATP) by analyzing the logs and detecting unusual behavior or suspicious activity that may lead to harmful attempts to get access to customer data.

Information protection is the fourth layer of the security strategy. Azure Synapse Analytics secures customer data by encrypting it in motion or in transit with Transport Layer Security (TLS). It always enforces encryption (SSL/TLS) for all connections so as to implement the concept of encryption-in-transit. That means all data "in-transit" between the client and the server is always encrypted for security reasons. Azure Synapse Analytics also supports the concept of encryption-at-rest. It means that all the data that is stored in the database for Azure Synapse Analytics is always encrypted using Transparent Data Encryption (TDE). There is an option for customers to manage an encryption through Azure Key Vault. That means that it supports the concept called Bring Your Own Key (BYOK) for TDE. Customers can manage their encryption key, including rotation of the key whenever required, on their own using Azure Key Vault.

Azure Synapse Analytics also supports a concept called always encrypted or encryption-in-use. That means that the data stored in a specific column of the database will always be encrypted for certain users like administrators who are not supposed to have access to certain types of information. This can easily be supported by Azure Synapse Analytics. Azure Synapse Analytics also has a feature called dynamic data masking that limits the exposure of sensitive data to certain privileged users only. What it means is that the data will be masked dynamically if it is queried by non-privileged users.

Microsoft has also introduced Azure Private Link to strengthen its security offerings for various Azure services, and that is applicable to Azure Synapse Analytics as well. In Azure Private Link, a secure connection is established based on consent flow. It insulates all the data that flows between Azure Synapse Analytics and its consumers, as all the data flows through the Microsoft network and not through the public internet. This is definitely a good move toward strengthening data security not just for Azure Synapse Analytics but for the rest of the other Azure services as well.

Summary

In this chapter, we have taken a small step in learning Azure Synapse Analytics. We started with defining Azure Synapse Analytics at a high level. We also covered a basic comparison of Azure Synapse Analytics and Azure SQL Data Warehouse. This was followed by a discussion of why you should learn Azure Synapse Analytics and what the current situation is for data roles related demand in the job market.

As part of learning any new technology or platform, you should always look at and learn some of its important features. That will create a solid conceptual foundation for you. We did exactly that here and looked at some of the primary features of Azure Synapse Analytics. We touched upon the unified data analytics experience, powerful data insights, HTAP, unlimited scale, security, privacy, compliance, and so on. Azure Synapse Analytics provides various functionalities through certain key service capabilities, so it is important to have a basic understanding of those service capabilities. As part of that, we discussed many of its core capabilities, which include the ability to explore data lakes easily, its tight integration with Apache Spark, its BI and AI integrations, the ability to author hybrid data pipelines with no-code or low-code options through the usage of Synapse Pipelines, workload management, its enterprise data warehousing capabilities, security, and more.

With this chapter, we have gained solid and detailed introductory knowledge about Azure Synapse Analytics. Based on this foundation, we are going to explore Azure Synapse Analytics in much more detail. Let us head toward the next chapter, in which we are going to explore Azure Synapse Analytics' architecture and core components.

CHAPTER 4

Architecture and Its Main Components

Azure Synapse Analytics is architecturally different than a traditional or modern data warehouse platform. It is a data lakehouse; hence, there are many architectural components that are different or new compared to a data warehouse. As discussed in previous chapters, Azure Synapse Analytics is a comprehensive data analytics platform that includes many tools and technologies.

In the previous chapter, we got a very detailed introduction to Azure Synapse Analytics. It gave us a good foundation, as we discussed the main features and key service capabilities that make Azure Synapse Analytics a promising and futuristic platform for any data analytics project. When you want to learn a new technology platform or tool, it is necessary to understand its core architecture and its main components, as that will instill in you deep knowledge and the ability to use said platform or tool in the most appropriate and efficient way so as to meet your core business requirements.

In this chapter, we are going to discuss Azure Synapse Analytics' architecture and its main components in more detailed manner. Since Azure Synapse Analytics consists of multiple tools and technologies of which a few are proprietary and a few are open source, its architecture can be a little complex and difficult to understand. As there are many components that interact with one another, the platform provides a comprehensive toolset with a nice GUI-based user interface to make it easy for end users to use it.

We will start with the overall high-level architecture of Azure Synapse Analytics and will also discuss the main components of its architecture in more detail by dedicating a separate section to each of its main components.

© Bhadresh Shiyal 2021
B. Shiyal, *Beginning Azure Synapse Analytics*, https://doi.org/10.1007/978-1-4842-7061-5_4

High-Level Architecture

Let us look at the high-level architecture of Azure Synapse Analytics (Figure 4-1).

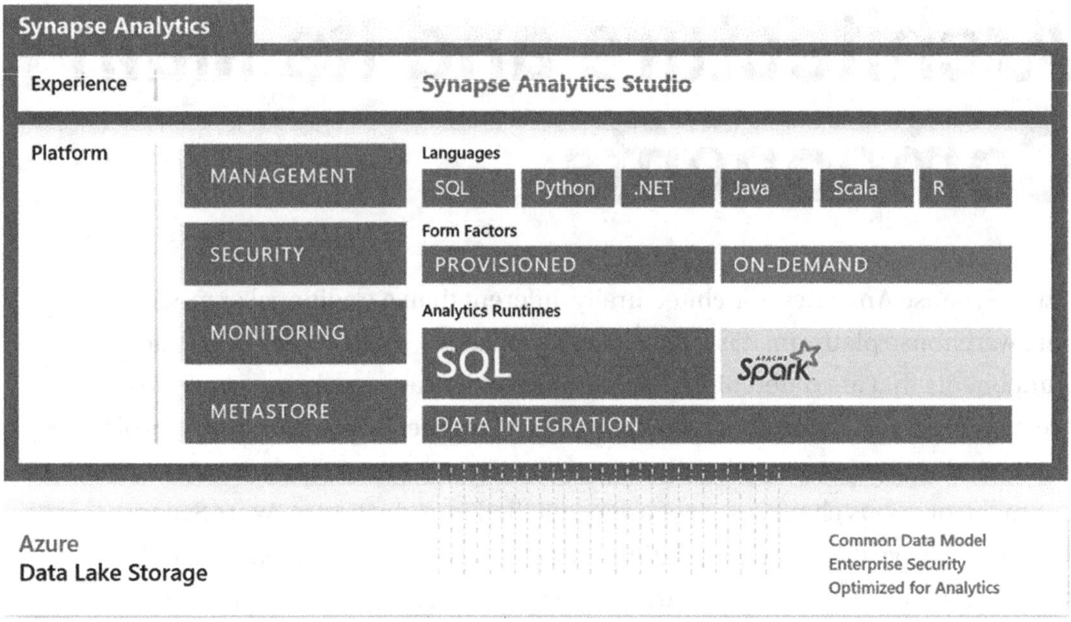

Figure 4-1. *Azure Synapse Analytics—high-level architecture. Source:* `https://docs.microsoft.com/en-us/learn/modules/introduction-azure-synapse-analytics/3-how-works`. *This figure is taken from Microsoft Learn Documentation. It is part of unit 3, named "How Azure Synapse Analytics Works" in a module named "Introduction to Azure Synapse Analytics" for the learning path named "Realize Integrated Analytical Solutions with Azure Synapse Analytics"*

We can divide the architecture of Azure Synapse Analytics into three different layers. These layers are the Experience layer, the Platform layer, and the Storage layer. Here, as part of the Storage layer, Azure Synapse Analytics supports Azure Data Lake Storage (ADLS) Gen2. We have discussed ADLS in previous chapters, so we are not going to discuss that here. However, we are going to discuss the remaining two layers in detail.

The Experience layer, which we can also recognize as the User Experience layer, is the very first layer in the overall architecture of Azure Synapse Analytics. This is the layer with which most of users are going to interact. It consists of a single but very powerful component called Synapse Studio, which is also known as Azure Synapse Analytics

Studio as well as Azure Synapse Studio. This is the top layer with which you will interact with the remaining two layers of the architecture. You can carry out almost all activities that are available within Azure Synapse Analytics using Synapse Studio. It is part of the Azure Synapse Workspace, which is a secured logical collaboration boundary in Azure Synapse Analytics. Synapse Studio gets provisioned automatically when you initially provision your Azure Synapse Workspace.

Synapse Studio is the go-to tool for anyone who wants to ingest, explore, analyze, or visualize the data stored in Azure Synapse Analytics. Whether a data engineer or data analyst or data scientist or database administrator, all will be able to carry out their tasks and activities using Synapse Studio. As the same tool is being used by everyone irrespective of their roles in the project, the level of collaboration increases within the team, which results in quicker rollout of new features in a product and saves time and effort as well.

The middle layer in the Azure Synapse Analytics architecture consists of multiple components. On top, it interacts with the Experience layer, while at the bottom it interacts with the Storage layer. The middle layer contains multiple components that are crucial for Azure Synapse Analytics as a platform; hence, we can call it the Platform layer. Two different analytics engines are at the core of the layer. These engines provide compute resources for Azure Synapse Analytics, and include Synapse SQL and Synapse Spark or Apache Spark. These analytics engines come in two different form factors in Azure Synapse Analytics: Dedicated or Provisioned and Serverless or On-Demand. Synapse SQL provides both the form factors, which means that we have Dedicated Synapse SQL as well as Serverless Synapse SQL or On-Demand Synapse SQL. However, when it comes to Apache Spark or Synapse Spark Engine, we have only one option, and that is Serverless Synapse Spark. In case of a Dedicated or Provisioned analytics engine, the compute resources are provisioned up front before it can be used, and so those are dedicated resources for those engines. However, in the case of a Serverless analytics engine, the compute resources are not provisioned up front, and so those are not dedicated or provisioned resources but rather get created on-demand as and when required.

Data integration and orchestration is also a core part of the Platform layer within the Azure Synapse Analytics architecture. It is responsible for integrating the disparate data sources and orchestrating the ETL/ELT jobs for your workload. This is the component in which Synapse Pipelines plays a crucial role. It is architecturally similar to Azure Data Factory. It provides various options to apply business logic and transformations on the data ingested from those data sources. This component opens up a lot of integration opportunities with third-party applications because it has around 80 different connectors, which are available out of the box to connect to those sources, copy the data, and transform it as per business requirements. This is a no-code solution that creates the data pipelines visually without you writing any code. Hence, Synapse Pipelines is an important component in Azure Synapse Analytics.

There are other components that are part of the Platform layer, including management, security, metastore, and monitoring components. Also, analytics engines and compute engines support the multiple languages in which you can write the code for your data analytics project. These components are tightly coupled with one another via the Azure Synapse workspace within your subscription. Security is of paramount importance for any architecture, and Azure Synapse Analytics provides a robust security mechanism. It allows you to have encryption of data at rest along with a second layer of encryption using customer-managed keys. The workspace-level configuration is available to enable data for double encryption. Apart from this, the platform also provides robust network security with multiple features to safeguard your data. Network security for Azure Synapse Analytics includes firewall rules, managed virtual networks, managed private endpoints, private links, and more. For authentication purposes, the platform is integrated with Azure Active Directory, which is the de facto standard for almost all Azure services. Along with AAD, it supports multi-factor authentication as well as SQL authentication. Azure Synapse Analytics also supports role-based access control (RBAC), which is the case with most of the other Azure services as well.

Synapse Link is also an important component added to Azure Synapse Analytics by Microsoft. Synapse Link creates a bridge between your operational data store and your analytical data store without you writing any code. Currently, it supports storing your operational data as well as analytical data in Azure Cosmos DB. Hence, Synapse Link makes it possible to generate an instant insight into your operational data as it gets synced automatically with your analytical data store in Azure Cosmos DB. We will discuss Synapse Link in a separate section later in this chapter.

As you have seen, the Platform layer in the Azure Synapse Analytics architecture contains many components, most of which play important roles within the overall Azure Synapse Analytics architecture. Synapse SQL, Synapse Spark, Synapse Pipelines, Synapse Studio, and Synapse Link are some of the important components, and we are going to discuss them in more detail in the upcoming sections of this chapter. These are the components you will have to deal with regularly while implementing any data analytics project in Azure Synapse Analytics.

Main Components of Architecture

As we saw in the high-level architecture review, Azure Synapse Analytics consists of multiple architectural components, each of which has its own unique importance within Azure Synapse Analytics. In this section, we are going to discuss the main components of the Azure Synapse Analytics architecture in a little more detail. All of these components will be discussed in much more detail in subsequent chapters, as there are dedicated chapters for all the main architectural components of Azure Synapse Analytics, which include Synapse SQL, Synapse Spark, Synapse Pipelines, Synapse Link, and Synapse Studio. We will start with Synapse SQL.

Synapse SQL

Let us take a deeper dive into Synapse SQL's architecture components. There are two important components in Synapse SQL: Compute layer and Storage layer. As we have discussed in previous chapters, the Compute layer is separate from or independent of the Storage layer. This enables us to scale the Compute resources as per our requirements independent of the data we have stored in Azure Synapse Analytics. Let us look at the Compute layer first, which will be followed by the Storage layer.

Compute Layer

There are two different options for the Compute layer in Synapse SQL, as shown in Figure 4-2. The first option is a Dedicated Synapse SQL Pool, and the second option is a Serverless Synapse SQL Pool. Let us examine the architecture of both options.

Dedicated Synapse SQL Pool

A Dedicated Synapse SQL Pool, as its name suggests, gets the computing power through dedicated compute resources. It consists of a well-known massively parallel processing (MPP) engine. For Dedicated Synapse SQL Pools, the data warehouse unit (DWU) is the unit of scale, which is being carried forward from Azure SQL Data Warehouse. It is an abstraction unit to measure the compute power for Dedicated Synapse SQL Pools. Dedicated Synapse SQL uses node-based architecture. There are two different types of nodes in its architecture. The first type of node is a control node, which is the single point of entry for any application or user. Once an application or user is connected to a control node in the Dedicated Synapse SQL Pool, they can issue T-SQL commands, which will then be sent to the MPP engine for distribution purposes. The engine will distribute the commands or queries to multiple compute nodes, which are the second type of node in the Azure Synapse SQL architecture.

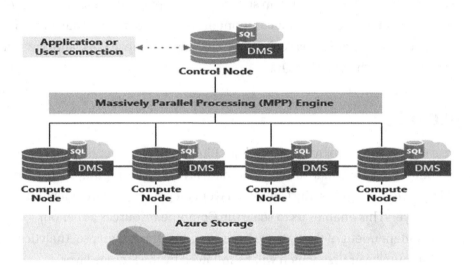

Figure 4-2. *Dedicated Synapse SQL architecture in Azure Synapse Analytics. Source: This figure is taken from Microsoft Documentation titled "Azure Synapse SQL Architecture" dated April 5, 2020.* `https://docs.microsoft.com/en-us/ azure/synapse-analytics/sql/overview-architecture`

The compute nodes are responsible for running your queries or commands in parallel to speed up the processing. Since it is distributed processing, in which multiple nodes process the data, there is a need to move data across the compute nodes, and for that purpose there is a system-level internal service. This service is known as DMS (Data Movement Service) and it helps to run the queries and commands in parallel and provides accurate results.

Serverless Synapse SQL Pool

A Serverless Synapse SQL Pool, as its name suggests, does not have dedicated computing power, unlike Dedicated Synapse SQL Pools. Based on your queries or commands, it will provision the computing resources for you behind the scenes. So, you need not worry about provisioning your computing resources up front, which is required for Dedicated Synapse SQL Pools. Serverless Synapse SQL Pools use a distributed query processing (DQP) engine, which is slightly different than the MPP engine being used by Dedicated Synapse SQL Pools. It also does not use the DWU concept, but rather relies on the size of the data being processed by your query for chargeback. Serverless Synapse SQL Pools also use node-based architecture. Here also you have two different types of nodes. The first type of node is a control node, which is the single point of entry for any application or user. Once an application or user is connected to the control node in a Serverless Synapse SQL Pool, they can issue T-SQL commands, which will then be sent to the DQP engine for distribution purposes (Figure 4-3).

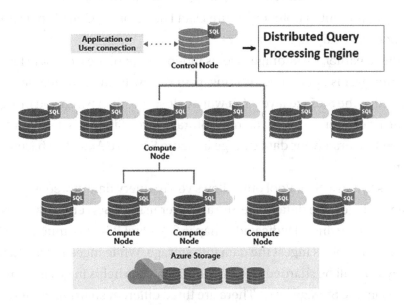

Figure 4-3. *Serverless Synapse SQL architecture in Azure Synapse Analytics. Source: This figure is taken from Microsoft Documentation titled "Azure Synapse SQL Architecture" dated April 5, 2020.* `https://docs.microsoft.com/en-us/ azure/synapse-analytics/sql/overview-architecture`

The DQP engine will be sure to optimize and orchestrate the distributed execution of the commands or queries sent by the user or an application. The DQP engine will split the queries into smaller queries and send those smaller queries to multiple compute nodes, the second type of node. Each small query sent to a compute node is known as a task, which can independently represent a distributed execution unit. As part of small query execution, it can read the data from storage and join results from other tasks. It can also group or sort the data received from other tasks.

Storage Layer

Storage is the second layer after the Compute layer in the Azure Synapse SQL architecture. Irrespective of which Compute layer option you decide to use, there needs to be a storage layer to store the processed data. You have two primary options for the Storage layer. Based on your requirements, you can decide to use either Azure Blob Storage or Azure Data Lake Storage Gen2; both of these options are available for both Compute layer options—Dedicated Synapse SQL Pools as well as Serverless Synapse SQL Pools. It is important to note that Azure Data Lake Storage Gen1 is not supported as a Storage layer in Azure Synapse Analytics.

As mentioned earlier, the Storage layer is independent of the Compute layer, so even if your Compute layer is not up and running, your data will still be available to you, as it is safely stored in the Storage layer. You will not lose any data when you bring down your Compute layer after storing data in the Storage layer. Since it is a separate layer, you are also charged separately for data storage in Azure Synapse Analytics based on your consumption.

A Serverless Synapse SQL Pool only allows you to query data stored in your Storage layer; you cannot ingest the data to the Storage layer using a Serverless Synapse SQL Pool. However, when using a Dedicated Synapse SQL Pool it is very much possible to query the data as well as ingest the data into storage. While ingesting the data into the Storage layer, it will be sharded into distributions. This helps in getting optimum performance from the Storage layer. There are three different sharding patterns available to you, which include Hash, Round Robin, and Replicate.

For Serverless Synapse SQL Pools, we have an additional option for the Storage layer—Azure Cosmos DB Analytical Store. However, this is only possible when you are using Azure Synapse Link. At the time of writing of this book, this feature was in preview. So, whenever it goes into general availability, a few things may change on this front.

Synapse Spark or Apache Spark

For any Big Data analytics applications, Apache Spark is considered as one of the de facto standards as it is a parallel processing framework. As we discussed in the previous chapter, Spark processes everything in-memory, and hence it always gives better performance than traditional Hadoop systems. Microsoft has done their own implementation of Apache Spark for Azure Synapse Analytics, which we refer to as Synapse Spark (Figure 4-4).

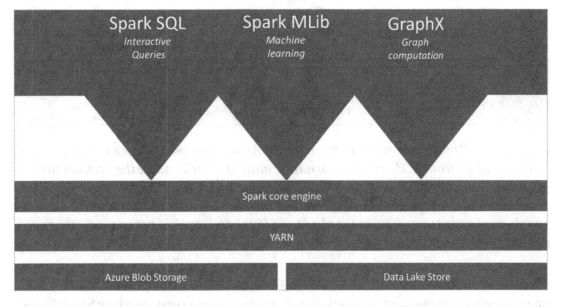

Figure 4-4. *Apache Spark or Synapse Spark in Azure Synapse Analytics. Source: This figure is taken from Microsoft Documentation titled "Apache Spark in Azure Synapse Analytics" dated April 5, 2020.* `https://docs.microsoft.com/en-us/ azure/synapse-analytics/spark/apache-spark-overview`

In Azure Synapse Analytics, it is very easy to create a Synapse Spark Pool. You will be able to provision the Spark Pools' clusters very quickly. These clusters are groups of computers that handle the execution of commands. Like Spark SQL, this follows a node-based architecture. There are two different types of nodes in Spark as well. However, their names are different in the case of Spark Pools. Here, the main node through which an application or user interacts with the Spark engine is known as a Driver node. As its name suggests, it drives the Spark engine and execution of commands issued to it by distributing them to another type of node within Spark, which is known as a Worker node (Figure 4-5).

Worker nodes are responsible for reading the data from data sources, processing it, and then writing the processed data back to storage. You can define the maximum number of nodes that your cluster can spin up for you. Spark also supports automatic scaling of nodes in which, based on the workload requirement, it will automatically spin up the additional Worker nodes up to the defined maximum limit and will automatically bring down the worker nodes when those are not required.

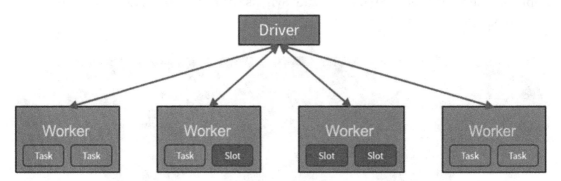

Figure 4-5. *High-level Spark architecture in Azure Synapse Analytics. Source: This image is taken from a Microsoft Learn unit named "Understand the architecture of Databricks Spark Cluster" under a module named "Spark Architecture Fundamentals."* https://docs.microsoft.com/en-us/learn/modules/ spark-architecture-fundamentals/2-understand-architecture-of-azure- databricks-spark-cluster

Azure Synapse Analytics provides a fully managed Spark Pool service. There are many benefits of using Synapse Spark. For example, you can create Synapse Spark Pools very easily using Azure Portal, Azure PowerShell, or .NET SDK for Synapse. The creation of a Spark node is also very fast, and it provides efficiency by bringing down the nodes when they are not required. To consume the compute power provided by Spark, it is very easy to use a notebook to write the code and execute it on the Spark cluster.

Spark Pool–based clusters are easily scalable. For automatic scaling, you can define the minimum and maximum nodes for your Spark Pool cluster, and it will take care of scaling up the number of worker nodes based on the runtime requirements of your workload. Similarly, it will scale down the number of worker nodes when they are not required. Thus, Spark Pool clusters are highly scalable and provide cost-efficiency as well. Spark Pool also allows you to use Azure Data Lake Storage Gen2 or Azure Blob Storage for storing the data. Spark Pool can be used to perform exploratory data analysis, machine learning, data visualization, and so forth by using various libraries that are available in the Spark ecosystem.

Synapse Pipelines

Synapse Pipelines is a cloud-based data integration and ETL/ELT service that is a core part of the Azure Synapse Analytics architecture. You can create data-driven workflows to move and transform data at scale using the Synapse Pipelines. You can build complex ETL or ELT processes to move and transform data using the GUI. It is a fully no-code, GUI-based service; hence, it allows you to develop the processes visually (Figure 4-6).

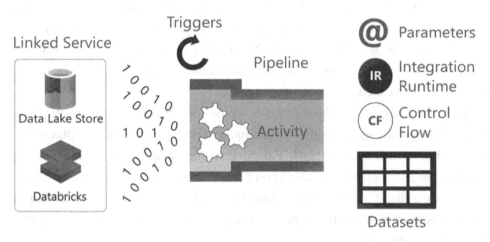

Figure 4-6. *Synapse Pipelines in Azure Synapse Analytics. Source: This image is taken from a video from a Microsoft Learn unit named "Understand Azure Data Factory Components" under the module named "Integrate Data with Azure Data Factory or Azure Synapse Pipelines."* `https://docs.microsoft.com/en-us/learn/ modules/data-integration-azure-data-factory/5-understand-components`

As far as its functionality and core components go, Synapse Pipelines derives almost everything from Azure Data Factory. However, the best part is that Synapse Pipelines is directly available within your Azure Synapse Analytics workspace inside Synapse Studio. Similar to Azure Data Factory, Synapse Pipelines consists of many core components. Let us discuss those components briefly one by one to understand their importance and functionality.

Synapse Pipelines allows you to connect multiple disparate data sources through a core component called linked services. In simple terms, linked services are connection strings that contain details about various parameters required in order to connect a specific data source using Synapse Pipelines. The linked services that allow you to connect various data sources are known as data store linked services. There is also another type of linked service called a compute linked service. That is used to define and provision the compute infrastructure on which you can execute pipeline workloads.

The dataset is another core component in Synapse Pipelines. It represents the data structure within a specific data source that you can connect to using the data store linked service. The datasets are used to provide data from data sources to another core component in Synapse Pipelines called an activity. You can include multiple activities together to apply any business logic or transformations to meet your business requirements; they will be applied to your dataset through activities. A logical grouping of activities creates another core component of Synapse Pipelines known as pipelines, from which the name of the Synapse Pipelines feature is derived. Pipelines can be executed on a time-based schedule, or you can define a trigger, which will decide when the pipeline will get executed.

Control flow is an orchestration of various pipeline activities that you have included in a pipeline. Sequencing, chaining, branching, and parameterizing your pipeline can all be achieved using control flow. Parameterization allows your pipeline to make dynamic decisions based on the values you pass to those parameters. The values of the parameters are utilized by various activities that you have included in your pipeline. A parameter is a key–value pair that is read-only and cannot be changed during the execution of the pipeline.

The last and most important core component of Synapse Pipelines is something called Integration Runtime (IR). As its name suggests, it is the runtime that will be used by the activities included in your pipeline. It is a simple bridge between data sources with which you connect via linked services and the activities that you want to perform on the datasets that you have derived from those data sources. Basically, IR provides compute infrastructure for execution of your data pipeline. Apart from Azure Integration Runtime, which comes with Synapse Pipelines by default, you have the option to set up your own integration runtime. It is known as Self-Hosted Integration Runtime (SHIR), and, as its name says, you will have to create a hosting environment on which the integration runtime will be installed and be responsible for providing compute power to your pipeline activities.

Thus, Synapse Pipelines is a core architectural component in Azure Synapse Analytics as it provides easy-to-use options to connect to various disparate data sources outside Azure as well. It provides a runtime to execute various business logic and transformations along with capabilities to orchestrate for an end-to-end data movement and transformation data pipeline experience without leaving Azure Synapse Studio. We will take a deeper look into Synapse Pipelines with a full chapter dedicated to it later in this book. Now, let us look at Synapse Studio in a little more detail.

Synapse Studio

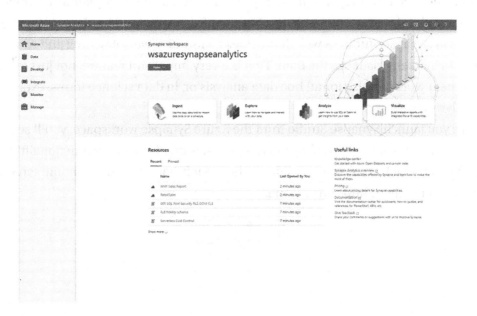

Figure 4-7. *Unified analytics experience through Synapse Studio. Source: The image is taken from the Microsoft Azure Synapse Analytics site under the link "Harness the power of a unified analytics experience."* `https://azure.microsoft.com/en-us/services/synapse-analytics/`

Synapse Studio is a one-stop shop for all your needs within Azure Synapse Analytics (Figure 4-7). It is a cloud-native web-based GUI tool that provides a central place from which you can carry out all kinds of tasks and activities for Azure Synapse Analytics. As we know by now, Azure Synapse Analytics includes many different tools and technologies, and it is extremely important to have a common central architectural component like Synapse Studio that can bind all these different tools and technologies together in the most user-friendly way.

With Azure Synapse Analytics, Microsoft has also introduced a new concept called a Synapse workspace. In order to use Azure Synapse Analytics, you will have to create a Synapse workspace first. A Synapse workspace provides a secured collaboration space for implementing enterprise-level cloud analytics projects in Azure. Once you create an Azure Synapse Analytics workspace in your Azure subscription, you will be able to launch and use Synapse Studio from the workspace itself. So, it is easy to get started with Synapse Studio as it does not have any separate provisioning or configuration steps once you have your Azure Synapse Analytics workspace created for you.

81

Synapse Studio also provides a notebook-based development experience, which is very much needed nowadays for any comprehensive data analytics platform. Ntract is an open source cloud-native notebook development framework and has been integrated by Microsoft in Synapse Studio as well. It provides an intuitive way of writing and running the code to instantly visualize the data. That is a very important feature, not just for real development work but also for ad hoc data analysis or, in data science terms, exploratory data analysis (EDA).

Once you launch Synapse Studio from the Azure Synapse workspace, you'll see a home screen on the launch itself. Synapse Studio has divided its core functionalities into six main hubs, which are directly visible on the top left-hand side of the home screen. Let us briefly discuss these hubs one by one.

The first hub is the Home hub, which is immediately visible when you launch Synapse Studio. In the center of the Home hub are links to ingest, explore, analyze, and visualize your data. These are basically shortcuts to various tools that are available in Azure Synapse Analytics. Next is the Data hub, which gives access to your Serverless Synapse SQL databases as well as Dedicated Synapse SQL databases. It also gives access to external data sources and other linked services you have created. The third hub is the Develop hub, using which you will be able to write SQL scripts, Synapse notebooks, reports, and so forth. This is where the real development work will happen, and all data roles including data engineers, data analysts, data scientists, and so on will access this hub for writing their code.

The Integrate hub is basically there to take you to the Synapse Pipelines interface. The look and feel are similar to Azure Data Factory, as Synapse Pipelines is architecturally similar to Azure Data Factory. This is an important hub from an integration point of view as you need not to go to Azure Data Factory for any of your data integration and orchestration needs. The Integrate hub provides all those options within Synapse Studio itself. The Monitor hub is meant to give you various options to monitor your pipeline runs, view status of IRs, view Synapse Spark jobs, and so forth, along with historical details about activities that happened in your workspace. The Manage hub provides options to manage Synapse SQL pools, Synapse Spark pools, linked services, integration runtimes, and more.

We are going to move on to Synapse Link in the next section. We have a dedicated chapter later in this book in which we are going to discuss Synapse Studio in more detail.

Synapse Link

We briefly discussed this component in previous chapters at a very high level. Here, we are going to go into a little more detail. First, let us do a quick recap. Synapse Link is a hybrid transactional and analytical processing (HTAP) system. It allows you to generate near-real-time analytics using the operational data stored in Azure Cosmos DB. Synapse Link provides a tighter and more seamless integration with Azure Cosmos DB. It provides no-ETL analytics as you don't need to write any ETL jobs to copy the data from the operational store to the analytical store for analytics purposes, as you can generate the analytics directly from the operational data in Cosmos DB. Rest assured that when you do this, it is not going to adversely impact the performance of your operational data store.

The Azure Cosmos DB analytical store is different than the traditional Azure Cosmos DB transactional or operational store. It is a fully isolated columnar store that can provide you with large-scale analytics against your Azure Cosmos DB opertaional or transacational store. As part of Synapse Link, there is an automatic sync that happens between the analytical store and the transactional store in Cosmos DB. This sync process will ensure that the performance of your Azure Cosmos DB transactional store is not impacted in any way. Another important difference to note between analytical and transactional stores is that the transactional data store in Azure Cosmos DB will be schema agnostic, while the analytical data store in Azure Cosmos DB will be schematized when syncing it from the transactional data store. Since there is no schema in the transactional store, you do not have to worry about index management in your Azure Cosmos DB for the transactional store. Schema and indexes are required for analytics, and hence those are implemented in Azure Cosmos DB for the analytical data store.

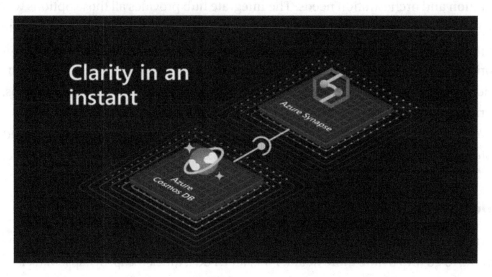

Figure 4-8. *Synapse Link for Cosmos DB in Azure Synapse Analytics. Source: The image is taken from the Microsoft Azure Synapse Analytics site under the link "Get instant clarity using the freshest operational data at all times."* `https://azure. microsoft.com/en-us/services/synapse-analytics/`

Synapse Link provides you with many advantages (Figure 4-8). No-ETL analytics is a big plus in favor of Synapse Link, as there is no need for any ETL jobs, as the data will be automatically synced within Cosmos DB from your transactional store to the analytical store. It saves a lot of effort that usually goes into building ETL jobs for any data engineering project. The syncing of data between transactional store and analytical store happens automatically in near real-time for those data that are required for analytics purposes. This gives you an opportunity to gain instant insights into your operational data as you don't have to build any ETL jobs, and hence there is no need to wait for them to finish before you can generate insights.

Azure Cosmos DB is optimized for scalability, elasticity, and performance, and you get all these benefits when you use Synapse Link for its intended purpose. Azure Cosmos DB is a truly globally available distributed database. That allows you to generate insights from your data, which is stored near to you in your region if you have opted for that. Azure Synapse Analytics gives two different options to query and analyze the data stored in your Azure Cosmos DB analytical store. It supports Synapse Spark Pool, which you can use to write your code so as to analyze data in different languages, including Scala, Python, Spark SQL, and C#. Apart from Synapse Spark, it also supports Serverless Synapse SQL Pool. However, you can use only T-SQL language with this, and you can also use the familiar BI tools.

Summary

In this chapter, we have discussed the Azure Synapse Analytics architecture and its main components in some detail. This chapter has given us a head start in understanding the technical details of Azure Synapse Analytics. Architecturally, Azure Synapse Analytics is a little complex to understand, as it involves multiple technical concepts, tools, and technologies. To make that simpler for you, we have divided the discussion into its main components and have discussed those components in separate sections.

Azure Synapse Analytics provides three different options when it comes to the compute engine. Synapse SQL provides two flavors: Dedicated Synapse SQL and Serverless Synapse SQL. We have discussed both flavors in some detail in separate sections. The third option is Synapse Spark, which is an Apache Spark implementation done by Microsoft specifically for Azure Synapse Analytics. As mentioned earlier, Spark is the de facto compute engine for most Big Data analytics projects, and for that reason Synapse Spark is one of the most exciting new features offered in Azure Synapse Analytics.

Later in the chapter, we discussed Synapse Pipelines, which provides you with the ability to design, develop, and deploy Azure Data Factory–based data pipelines. Architecturally, Synapse Pipelines is like Azure Data Factory. However, there are some minor differences, and Synapse Pipelines does not offer all the features offered by Azure Data Factory. Synapse Pipelines is directly integrated with Synapse Studio. You get all the familiar components from Azure Data Factory, like linked services, datasets, pipelines, triggers, integration runtime, and so on in Synapse Pipelines as well, and all these features are available in Synapse Studio itself, so you need not leave the Synapse workspace to carry out any Synapse Pipelines–related activities.

Toward the end of the chapter, we discussed a futuristic and advanced architectural component of Azure Synapse Analytics, which is Synapse Link. Hybrid transactional and analytical processing (HTAP) is the future of data analytics and provides a no-ETL analytics solution that is cost effective and easy to implement. We discussed how Azure

Cosmos DB for transactional stores can sync with Azure Cosmos DB for analytical stores. This provides instant insights into operational or transactional data stored in Cosmos DB. Synapse Link bridges the gap between your operational and analytical data stores and saves you a lot of effort, as you will not be required to build ETL jobs if you use Synapse Link in your project.

This chapter has given a somewhat detailed understanding of the Azure Synapse Analytics architecture and its main components. As mentioned earlier, in subsequent chapters we are going to discuss all these components in much more detail, as we have dedicated chapters for most of these components. In the next chapter, we are going to start this journey with Synapse SQL.

CHAPTER 5

Synapse SQL

Azure Synapse Analytics consists of multiple architectural components along with many tools and technologies. In the previous chapter, we discussed the Azure Synapse Analytics architecture and its main components in detail. During that discussion, we also briefly examined Synapse SQL as one of the important architectural components. In this chapter, we are going to discuss Synapse SQL in much more detail.

Synapse SQL is a core component of Azure Synapse Analytics, as it offers two different node-based architectures: Dedicated Synapse SQL and Serverless Synapse SQL. Both have slightly different internal workings, as well as a couple of common components. We will dive deeper into this, and will also cover massively parallel processing (MPP) and distributed query processing (DQP) engines in a little more detail so as to understand how each of these engines works internally. Both engines use a Control node and one or more Compute nodes. We will discuss what work these nodes carry out for both the engines.

While storing data to storage, it is necessary to shard the data using any of the sharding or data distribution patterns for improved performance. You can distribute your data using either round robin distribution or hash-based distribution. Alternatively, you can also consider replicating the full tables. We will discuss all these patterns in detail in this chapter.

We will also look at Dedicated Synapse SQL Pools and Serverless Synapse SQL Pools. Both pools have many things in common, and many things are different as well. It is necessary to compare the features of these two pools in detail. We will carry out the comparisons by grouping the various features into feature groups, like database object type, query language, tool support, storage options, data formats, and security.

© Bhadresh Shiyal 2021
B. Shiyal, *Beginning Azure Synapse Analytics*, https://doi.org/10.1007/978-1-4842-7061-5_5

Synapse SQL consumes some of the resources from Azure services. These resources decide how you scale your Synapse SQL architecture when required. Similarly, the resource consumption pattern will also decide how your billing will work. Dedicated Synapse SQL and Serverless Synapse SQL have different resource consumption models. We will discuss resource consumption models for both in detail before moving on to discuss some of the best practices for Synapse SQL. Best practices help us to implement the technology in the right manner the very first time and save a lot of time and effort, as these practices are proven and mostly work fine. They help to resolve or troubleshoot any technical implementation-level details. Synapse SQL offers two pools, and it makes sense to know the best practices for implementation for both of these pools.

Toward the end of the chapter, we are going to discuss some of the most common activities that you should know how to carry out for Synapse SQL. We will discuss step-by-step how to create a Dedicated Synapse SQL Pool, how to create a Serverless Synapse SQL Pool, how to load using the COPY statement, and how to ingest data into Azure Data Lake Storage (ADLS) Gen2.

Synapse SQL Architecture Components

We looked at the architecture for Synapse SQL at a high level in Chapter 4, where we discussed the overall Azure Synapse Analytics architecture. There are two different options available to you when it comes to Synapse SQL architecture. These two options are Dedicated Synapse SQL Pool and Serverless Synapse SQL Pool. These two are also known as Provisioned Synapse SQL Pool and On-Demand Synapse SQL Pool. Architecturally, they are similar to each other, but there are some subtle differences as well. In both architectures, the Control node, Compute node, and Azure storage are common components. Even though both use different types of engines, in the end both are node-based architectures.

In Figure 5-1, you can see the Dedicated Synapse SQL architecture on the left-hand side and the Serverless Synapse SQL architecture on the right-hand side. Since both are node-based architecture, you can see easily that there are two different types of nodes used in both architecture diagrams. The very first node is the Control node, and the rest are Compute nodes. In both cases, the entry point for any user, application, or connection will always be the Control node. There is no need to access the Compute nodes directly. That access will always be only via the Control node.

Dedicated SQL pool

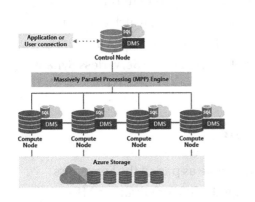

Serverless SQL pool

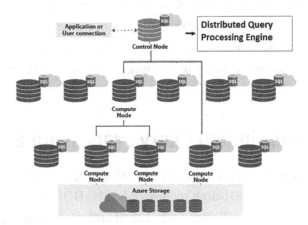

Figure 5-1. Synapse SQL architectures. Source: This image is taken from Microsoft Documentation titled "Azure Synapse SQL Architecture" under the sub-section named "Synapse SQL Architecture Components" dated April 15, 2020. `https://docs.microsoft.com/en-us/azure/synapse-analytics/sql/overview-architecture`

Since each architecture uses a different engine, let us discuss both of those engines in a little more detail here. Let us start with the massively parallel processing (MPP) engine.

Massively Parallel Processing Engine

This is one of the most widely used processing engines for data warehousing systems. Azure SQL Data Warehouse used the massively parallel processing engine and was pretty successful in supporting Big Data workloads as well. The same design from Azure SQL Data Warehouse is being carried forward in Dedicated Synapse SQL. Opposite to this, there is a concept called symmetric multi-processing (SMP). In former days, prior to MPP came into existence, SMP was widely used. It used multiple processors that were tightly coupled and shared resources like OS, memory, I/O, and so forth, which were directly connected to a common bus. Since this was not a highly scalable architecture, MPP came into the picture. It is more scalable than the SMP engine and so can support Big Data analytics workloads very easily. The MPP engine follows a node-based architecture and requires a Control node and multiple Compute nodes for data processing.

The MPP engine also requires a common data movement service (DMS) in the Control node as well as in each of the Compute nodes that are part of the engine. This service is essential in order to move data from one node to another, which allows us to take advantage of parallel processing at a massive scale. Now, let us look at the distributed query processing engine, which is available as part of the Serverless Synapse SQL architecture.

Distributed Query Processing Engine

The distributed query processing (DQP) engine is available as part of the Serverless Synapse SQL architecture only. The DQP engine is slightly different than the MPP engine. It still uses node-based architecture and will have both a Control node and Compute nodes. However, it does not have data movement service (DMS) as part of its architecture. The DQP engine's primary responsibilities are to optimize and orchestrate the distributed execution of the commands or queries sent by the user or an application. Once it receives a query, it splits that query into multiple smaller queries and sends those smaller queries to multiple different Compute nodes. Those smaller queries are known as tasks. To complete the tasks, the engine can read the data from storage. Later on, it can join results from other tasks and can also group or sort the data.

Compared to the MPP engine, the DQP engine is a query-only engine, as its name suggests. That means that you cannot use the DQP engine to ingest new data into your storage or update any data that is already there on your storage. This is meant for ad hoc querying only. This engine scales the nodes automatically based on the workload you provide to it.

We have seen that Synapse SQL architecture contains multiple components. The Control node, Compute node, data movement services, distributions, and storage are the most important components of the Synapse SQL architecture. Now, let us jump into discussing these components one by one. Let us start with the Control node.

Control Node

Both Dedicated Synapse SQL and Serverless Synapse SQL use node-based architecture, and both have a Control node and generally more than one Compute nodes as part of their architecture. In both architectures, the Control node is considered the only entry point for any user, application, or connection. The Control node plays a crucial role in controlling whole processing or querying exercises in both architectures. That is why we say that the Control node plays the role of the brain in both architectures.

In Dedicated Synapse SQL architecture, the Control node is responsible for running the massively parallel processing engine on itself. Here, once the query is issued to the Control node, it will first optimize the submitted query and then later will coordinate the execution of the queries in parallel. The Control node also uses the data movement service (DMS) running on it to distribute the queries across multiple compute nodes in the most optimum way possible.

In Serverless Synapse SQL architecture, the Control node is responsible for running the distributed query processing engine on itself. Similar to the Dedicated Synapse SQL architecture, once the Control node receives the query, it divides that query into multiple smaller queries. This is required in order to get the benefits of having multiple Compute nodes available for processing the submitted query in parallel; this gets the query results as fast as possible. However, there is one notable difference here, and that is the data movement service (DMS), which is not available to the Control node or Compute nodes in the Serverless Synapse SQL architecture.

Compute Nodes

The Compute nodes are present in both architectures. These Compute nodes provide the real processing or computing power to both engines.

In the Dedicated Synapse SQL architecture, the Control node distributes smaller queries to the available Compute nodes. You can have a minimum of one to a maximum of sixty Compute nodes in your Dedicated Synapse SQL Pool. So, based on available Compute nodes, the Control node will distribute the query workload to that many smaller queries so that each Compute node gets something to execute in parallel to the other Compute nodes. Each Compute node will also have its independent internal data movement service (DMS) on it. This helps the Compute nodes to move the data from one node to another node within the pool.

In the Serverless Synapse SQL architecture, each Compute node gets assigned a task by the Control node. The Control node uses the distributed query processing engine to distribute the smaller queries generated by it to various Compute nodes. Additionally, as it is a serverless architecture, the Compute nodes get provisioned on the fly, and you need not worry about its provisioning. Based on the query that the Control node has received, it will decide how many different Compute nodes it will require to execute that query in the most optimum possible way. Generally, the Control node will also assign a set of files from which each Compute node can query and retrieve the data. The best part about Compute nodes in the Serverless Synapse SQL architecture is that the scaling happens automatically based on your query workload. The engine is intelligent enough to decide how many Compute nodes it will require in order to execute a particular task, and accordingly it will make that many Compute nodes available for parallel query execution.

Data Movement Service

The data movement service (DMS) is the internal standalone and independent service that is installed on each of the Compute nodes and the Control node in the Dedicated Synapse SQL architecture. The DMS is the data transport technology, as its name suggests. It is available in the Dedicated Synapse SQL architecture and is not present in the Serverless Synapse SQL architecture. Microsoft made significant overall improvements in the DMS from Azure Data Warehouse Gen1 to Azure SQL Data Warehouse Gen2. The improved DMS is carried forward in the Dedicated Synapse SQL architecture as well.

The data movement service's primary responsibility is to coordinate data movement between the Compute nodes. In a distributed engine, it is not always possible that all the data required by a Compute node is readily available in that same node. There are many situations in which one Compute node will have to fetch data from another Compute node in order to complete the query execution assigned to a particular Compute node. The DMS is required to ensure that the parallel queries that are getting executed in different Compute nodes return accurate results at the end. Therefore, it is the responsibility of the DMS to make available the right data to the right location at the right time during the query execution across multiple Compute nodes.

Distribution

The distribution is an important architectural component that helps us to understand how Dedicated Synapse SQL's massively parallel processing (MPP) engine internally distributes the data for processing across the available Compute nodes. The basic unit of data storage and data processing for queries getting executed in parallel on the distributed data in the Dedicated Synapse SQL is known as a distribution.

Once the query is submitted to the Control node, which is the entry point for any queries submitted by any application, user, or connection, the Control node divides the submitted query into 60 smaller queries so that they can be executed in parallel on available data distributions. Each of these 60 smaller queries will run on one of the data distributions. A Compute node can manage only one or more than one data distribution. Let us assume that we have the maximum number of Compute nodes available to us, which is 60 by default. In this case, each Compute node will have just one data distribution on it, and each will execute one of the smaller queries on that data distribution, which is available on the Compute node. However, if we assume that there are only 30 Compute nodes available to us, then in that case, each Compute node will have a minimum of two data distributions assigned to it. Similarly, if we assume that there is only one Compute node available to us, then in that case all 60 data distributions will be available in that single Compute node.

When you ingest data into Azure storage using the Dedicated Synapse SQL Pool, it will shard the data into distributions to get the best system performance. You have three different sharding patterns available to you, which correspond to the same number of different distribution patterns. Therefore, there are mainly three types of distribution patterns available to use, which allows us to distribute the data in three different ways to meet different data scenarios. These are as mentioned in Figure 5-2.

Hash Distributed	Round Robin	Replicated
Distributed using hashing algorithm	Distributed evenly but randomly	All data present on every node
Equal values hash to same distribution	Does not require knowledge about data or queries	Simplifies many query plans and reduces data movement
Optimal for large fact tables	Optimal for large tables without a good hash column or varied queries	Best for small lookup tables

Figure 5-2. *Data distribution patterns*

These distribution patterns are applied when you ingest the data in the table. Let us examine each of them.

Hash Distribution

When the Dedicated Synapse SQL Pool uses a hash function to deterministically assign each row to one distribution, then it is known as hash distribution. The tables created using such a hash function are known as hash-distributed tables. In these types of tables, one column will be designated as a distribution column, and hash distribution will apply the hash function to the values in the designated distribution column to ensure that one row gets assigned to one distribution only.

Here, which column is to be picked up as the distribution column is a very important decision. In order to get the best performance, you should carefully study the values of various columns. You should particularly pay attention to the distinctness of the values as well as any skewed data. The important characteristic to rememeber for hash distribution is that each row in the hash-distributed table belongs to one and only one data distribution, and in no situation can it be part of more than one distribution (Figure 5-3).

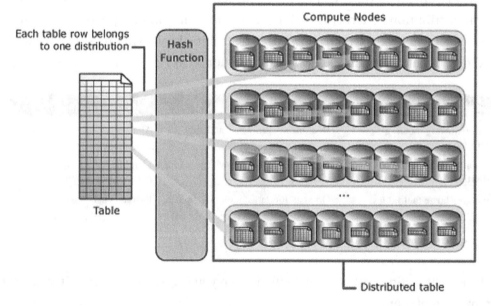

Figure 5-3. *Hash-distributed table in Dedicated Synapse SQL. Source: This image is taken from Microsoft Documentation titled "Azure Synapse SQL Architecture" under a sub-section named "Hash-distributed tables" dated April 15, 2020.*
https://docs.microsoft.com/en-us/azure/synapse-analytics/sql/
overview-architecture#hash-distributed-tables

Hash-distributed tables make more sense when you have large fact tables that you want to distribute properly to get optimum performance. As per the general guidance, hash-distributed tables should be used when you have a table size greater than 2 GB and you are expecting a significant number of insert, update, and delete operations on that table.

Round-Robin Distribution

When the Dedicated Synapse SQL Pool distributes the data in a round-robin or random manner, but evenly, then it is called round-robin distribution. This is one of the easiest distributions to apply as it does not require any prior knowledge of any values in columns, since it does not depend on any column values for distribution purposes. It is a good fit for staging tables, as it can load the data very fast. However, it is not possible to get any additional optimization while querying the data stored in round robin–distributed tables.

If you are looking for optimized query performance, then you should look for hash distribution rather than round-robin distribution. You should also avoid round-robin distribution if your table is going to be used in joins, as it will not give good performance. This is so because while joining the round robin–distributed table, the system will have to shuffle the data to complete the join obligations. Hence, it will consume more time to complete the joins and give you output after the additional shuffle time. Round-robin distribution is generally used when you cannot find a good candidate column for applying hash functions. Simiarly, when you know that your queries on the tables are going to be varying each time, then it is a good idea to go with round-robin distribution. As a rule of thumb, you can remember that round-robin distribution is good for data loading but is not the optimum solution while querying the data. Hence, you need to decide very carefully about when to use hash distribution and when to use round-robin distribution based on your specific use case; otherwise, your distribution can impact your performance adversely.

Replication-based Distribution

When the Dedicated Synapse SQL Pool replicates all the data of a table in each Compute node, then it is called replication-based distribution, and the tables so created are known as replicated tables (Figure 5-4).

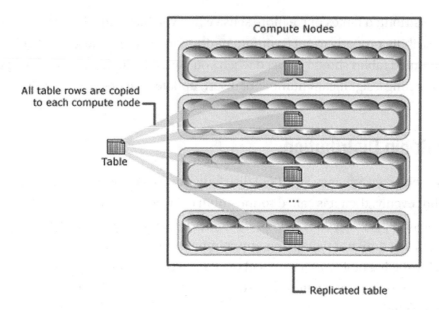

Figure 5-4. *Replication-based distribution in Dedicated Synapse SQL. Source: This image is taken from Microsoft Documentation titled "Azure Synapse SQL Architecture" under a sub-section named "Hash-distributed tables" dated April 15, 2020.* `https://docs.microsoft.com/en-us/azure/synapse-analytics/sql/overview-architecture#replicated-tables`

Replicated tables are really useful when your table size is smaller. The general rule of thumb is to replicate tables that are less than 2 GB in size. However, in certain situations you can decide to replicate tables even larger than 2 GB in each of the Compute nodes. Generally, dimension tables are smaller in size and are good candidates for replication-based distribution. The replicated tables will always provide better query performance for smaller tables.

Let us assume that you need to join a couple of dimension tables with your fact table to produce some analytics reports. Generally, your fact table will be larger in size, while your dimension tables will be smaller in size. Now, if you don't replicate your smaller dimension tables, then your Compute nodes will have to get the necessary data for the dimension tables from another Compute node through the data movement service (DMS), which will impact the performance adversely. However, if you have replicated your smaller dimension tables in each of your Compute nodes, then while joining the distributed data of the fact table with the dimension tables each Compute node will have the required dimension tables' related data available within itself. This results in improved performance for your joins. There is a slight overhead of copying all data to all the Compute nodes in order to get the benefits of replicated tables. So, you should weigh your options adequately before zeroing in on replicated tables.

In Apache Spark, there is the concept of the broadcast join, in which you copy all the data of a smaller table to each worker node to improve the join performance by avoiding the shuffling of data among worker nodes. A replicated table is exactly the same fundamental concept and is available in Dedicated Synapse SQL.

Azure Storage

Whether it's Dedicated Synapse SQL or Serverless Synapse SQL, both architectures require a storage layer to store the data persistently. Azure storage is used in both cases. As you know, the Dedicated Synapse SQL Pool allows you to query and ingest the data into Azure storage, while the Serverless Synapse SQL Pool only allows you to query the data; it does not allow you to ingest or update the data. Therefore, when the Dedicated Synapse SQL Pool is used, the data being stored in Azure storage will be sharded by using various distribution patterns, which we have just discussed.

One important point to note here is that the Storage layer is always separate from the Compute layer, and therefore both are charged for separately. As your storage is not part of your Compute layer, you can bring down your Compute layer without impacting your data that is already stored in Azure storage. As the Compute and Storage layers are decoupled completely, this facility will allow you to save some infrastructure usage cost.

Dedicated or Provisioned Synapse SQL Pool

In order to bring two different worlds—namely Big Data analytics (BDA) and enterprise data warehouse (EDW)—together, Microsoft introduced Azure Synapse Analytics. As a first step, Microsoft rebranded and included Azure SQL Data Warehouse in Azure Synapse Analytics so that it could take care of the EDW world within it. Therefore, Dedicated or Provisioned Synapse SQL Pools became part of Azure Synapse Analytics. The Dedicated Synapse SQL Pool is a representation of a collection of analytics-specific resources being used by the Dedicated Synapse SQL architecture (Figure 5-5).

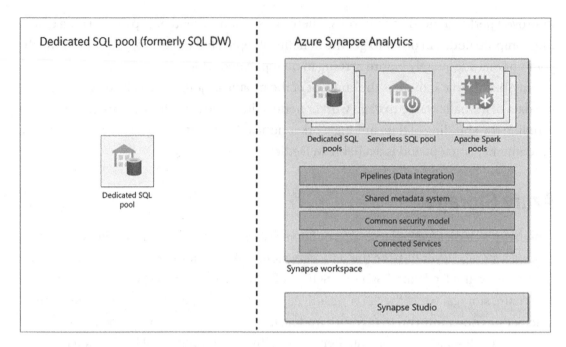

Figure 5-5. *Dedicated or Provisioned Synapse SQL Pool (formerly Azure SQL Data Warehouse). Source: This image is taken from Microsoft Documentation titled "What is dedicated SQL Pool (formerly SQL DW) in Azure Synapse Analytics?" dated November 4, 2019.* `https://docs.microsoft.com/en-us/ azure/synapse-analytics/sql-data-warehouse/sql-data-warehouse- overview-what-is?`

After creating the Dedicated Synapse SQL Pool, you will be able to ingest the data using PolyBase, which uses a simple T-SQL-based query to load the data quickly. Once the data is loaded, you will be able to run a highly performant data analytics workload on that data using the massively parallel processing (MPP) engine. Here, the data is stored in relational tables with columnar storage, which significantly reduces data storage costs and improves the performance of the query execution. As a result of the MPP engine, which Dedicated Synapse SQL Pool uses internally, there is a significantly noteworthy performance improvement over any traditional database management system (DBMS).

It is also very easy to create or provision a Dedicated Synapse SQL Pool once you have created your Azure Synapse Analytics workspace in your Azure subscription. We will look at how to create or provision a new Dedicated Synapse SQL Pool using Azure Portal later in this chapter, when we discuss step-wise details.

In addition to the Dedicated Synapse SQL Pool, there is a second option that is also part of the Synapse SQL architecture: Serverless Synapse SQL Pool. Let's take a look.

Serverless or On-Demand Synapse SQL Pool

We already discussed the Serverless Synapse SQL architecture earlier in this chapter. As part of this architecture, you get an already provisioned end point that is ready for use—a Serverless Synapse SQL Pool. This is exactly opposite to the Dedicated Synapse SQL Pool. As you don't have to provision or dedicate any specific infrastructure up front in order to start using this option, it is known as a Serverless or On-Demand Synapse SQL Pool. This pool is immediately available to you once you create an Azure Synapse Analytics workspace in your Azure subscription. You are ready to start using Serverless Synapse SQL Pool.

Since the Compute nodes are provisioned for you by Azure automatically based on your query workload, the charges are not based on the number of Compute nodes being used in the Serverless Synapse SQL Pool. Instead, you will be charged based on the size of the data being used for your query execution. So, there will be charges per terabyte of data processed for your query execution. This makes it pay-as-you-go (PAYG) in the truest sense.

A Serverless Synapse SQL Pool is a data querying service that allows you to query the data stored in your Azure Data Lake Storage. You have two options to query your data lake using a Serverless Synapse SQL Pool. The first option is to use the very common T-SQL syntax to query your data lake without loading or copying the data into a specialized data store. This is one of the greatest advantages of using Serverless Synapse SQL Pools. In-place instant querying can allow you to generate quicker insights into data that is already present in your data lake. Generally, you will use Azure Synapse Studio for this option. Another option is to use various other BI- and T-SQL-based querying tools, which include SQL Server Management Studio (SSMS), Power BI, and Azure Data Studio.

A Serverless Synapse SQL Pool is useful to all data professionals, including data engineers, data scientists, data analysts, and Power BI professionals. It is useful in many different situations. For example, data engineers can use a Serverless Synapse SQL Pool to explore data from a data lake and use the findings to optimize data transformations, which can allow them to build more efficient and simplified data transformation pipelines. This is possible because Serverless Synapse SQL Pools support reading and exploring data stored in various formats, like Parquet, CSV, JSON, and so on. Similarly, data scientists can benefit by using it to quickly carry out exploratory data analysis (EDA) on the data stored in the data lake. It supports OPERROWSET, which reads the data from a remote data source like files and returns the content as set of rows as well as automatic schema inference, which are really beneficial to data scientists as part of their initial exploratory work. This work is crucial in order to determine the right type of algorithm to be applied to the data stored in data lake storage.

Synapse SQL Feature Comparison

As we have both Dedicated and Serverless Synapse SQL options available to us, it is necessary to understand which features are supported and not supported in each. For ease of comparison, the features are grouped together in various categories like Database Object Types (Table 5-1), Query Language (Table 5-2), Security (Table 5-3), Tools (Table 5-4), Storage (Table 5-5), and Data Formats (Table 5-6). We will use tabular formats to compare various features.

Database Object Types

Table 5-1. *Database Object Types Feature Comparison*

Object Types	Dedicated	Serverless
Tables	Yes	No, serverless model can query only external data placed on Azure storage
Views	Yes. Views can use query language elements that are available in dedicated model.	Yes. Views can use query language elements that are available in serverless model.
Schemas	Yes	Yes
Temporary tables	Yes	No
Procedures	Yes	Yes
Functions	Yes	Yes, only inline table-valued functions
Triggers	No	No
External tables	Yes. See supported data formats.	Yes. See supported data formats.

(continued)

Table 5-1. (*continued*)

Object Types	Dedicated	Serverless
Caching queries	Yes, multiple forms (SSD-based caching, in-memory, result-set caching). In addition, Materialized Views are supported	No
Table variables	No, use temporary tables	No
Table distribution	Yes	No
Table indexes	Yes	No
Table partitions	Yes	No
Statistics	Yes	Yes
Workload management, resource classes, and concurrency control	Yes	No
Cost control	Yes, using scale-up and scale-down actions	Yes, using the Azure Portal or T-SQL procedure

There are some key differences between Dedicated Synapse SQL Pool and Serverless Synapse SQL Pool for database object type support. You can see those difference clearly if you pay attention to all those database object types for which Serverless Synapse SQL Pool reads "No" in Table 5-1. That means that those object types are not supported for Serverless Synapse SQL. In particular, tables, table distributions, table indexes, and table partitions are not supported in Serverless Synapse SQL. Similarly, workload management, resource classes, and concurrency control are not supported in Serverless Synapse SQL. There is nothing in the database object type category that is supported in Serverless Synapse SQL and that is not supported in Dedicated Synapse SQL. Hence, at least in case of supportability for various database object types, Dedicated Synapse SQL has the upper hand over Serverless Synapse SQL.

Query Language

Table 5-2. *Query Language Feature Comparison*

	Dedicated	Serverless
`SELECT` statement	Yes. Transact-SQL query clauses FOR XML/FOR JSON, and MATCH are not supported.	Yes. Transact-SQL query clauses FOR XML, MATCH, PREDICT, and query hints are not supported.
`INSERT` statement	Yes	No
`UPDATE` statement	Yes	No
`DELETE` statement	Yes	No
`MERGE` statement	Yes (preview)	No
Transactions	Yes	Yes, applicable on metadata objects
Labels	Yes	No
Data load	Yes. Preferred utility is COPY statement, but the system supports both BULK load (BCP) and CETAS for data loading.	No
Data export	Yes. Using CETAS.	Yes. Using CETAS.
Types	Yes, all Transact-SQL types except cursor, hierarchyid, ntext, text, and image, rowversion, Spatial Types, sql_variant, and xml	Yes, all Transact-SQL types except cursor, hierarchyid, ntext, text, and image, rowversion, Spatial Types, sql_variant, xml, and Table type
Cross-database queries	No	Yes, including USE statement.
Built-in functions (analysis)	Yes, all Transact-SQL analytic, conversion, date and time, logical, mathematical functions, except CHOOSE, and PARSE	Yes, all Transact-SQL analytic, conversion, date and time, logical, mathematical functions

(continued)

Table 5-2. (*continued*)

	Dedicated	Serverless
Built-in functions (text)	Yes. All Transact-SQL String, JSON, and Collation functions, except STRING_ ESCAPE and TRANSLATE	Yes. All Transact-SQL String, JSON, and Collation functions
Built-in table-value functions	Yes, Transact-SQL Rowset functions, except OPENXML, OPENDATASOURCE, OPENQUERY, and OPENROWSET	Yes, Transact-SQL Rowset functions, except OPENXML, OPENDATASOURCE, and OPENQUERY
Aggregates	Transact-SQL built-in aggregates except CHECKSUM_AGG, and GROUPING_ID	Transact-SQL built-in aggregates
Operators	Yes, all Transact-SQL operators except !> and !<	Yes, all Transact-SQL operators
Control of flow	Yes. All Transact-SQL Control-of-flow statement except CONTINUE, GOTO, RETURN, USE, and WAITFOR	Yes. All Transact-SQL Control-of-flow statement SELECT query in WHILE (...) condition
DDL statements (CREATE, ALTER, DROP)	Yes. All Transact-SQL DDL statements applicable to the supported object types	Yes. All Transact-SQL DDL statements applicable to the supported object types

Security

Table 5-3. *Security Feature Comparison*

	Dedicated	Serverless
Logins	N/A (only contained users are supported in databases)	Yes
Users	N/A (only contained users are supported in databases)	Yes
Contained users	Yes. Note: Only one Azure AD user can be unrestricted admin	No

(*continued*)

Table 5-3. (*continued*)

	Dedicated	Serverless
SQL username/password authentication	Yes	Yes
Azure Active Directory (Azure AD) authentication	Yes, Azure AD users	Yes, Azure AD logins and users
Storage Azure Active Directory (Azure AD) passthrough authentication	Yes	Yes
Storage SAS token authentication	No	Yes, using DATABASE SCOPED CREDENTIAL in EXTERNAL DATA SOURCE or instance-level CREDENTIAL
Storage access key authentication	Yes, using DATABASE SCOPED CREDENTIAL in EXTERNAL DATA SOURCE	No
Storage managed identity authentication	Yes, using Managed Service Identity Credential	Yes, using Managed Identity credential
Storage application identity authentication	Yes	No
Permissions— object-level	Yes, including ability to GRANT, DENY, and REVOKE permissions to users	Yes, including ability to GRANT, DENY, and REVOKE permissions to users/logins on the system objects that are supported
Permissions—schema-level	Yes, including ability to GRANT, DENY, and REVOKE permissions to users/logins on the schema	Yes, including ability to GRANT, DENY, and REVOKE permissions to users/logins on the schema
Permissions—database-level	Yes	Yes
Permissions—server-level	No	Yes, sysadmin and other server-roles are supported

(*continued*)

Table 5-3. (*continued*)

	Dedicated	Serverless
Permissions—column-level	Yes	Yes
Roles/groups	Yes (database scoped)	Yes (both server and database scoped)
Security & identity functions	Some Transact-SQL security functions and operators: CURRENT_USER, HAS_DBACCESS, IS_MEMBER, IS_ROLEMEMBER, SESSION_USER, SUSER_NAME, SUSER_SNAME, SYSTEM_USER, USER, USER_NAME, EXECUTE AS, OPEN/CLOSE MASTER KEY	Some Transact-SQL security functions and operators: CURRENT_USER, HAS_DBACCESS, HAS_PERMS_BY_NAME, IS_MEMBER' , ' IS_ROLEMEMBER, IS_SRVROLEMEMBER, SESSION_USER, SESSION_CONTEXT, SUSER_NAME, SUSER_SNAME, SYSTEM_USER, USER, USER_NAME, EXECUTE AS, and REVERT. Security functions cannot be used to query external data (store the result in variable that can be used in the query).
DATABASE SCOPED CREDENTIAL	Yes	Yes
SERVER SCOPED CREDENTIAL	No	Yes
Row-level security	Yes	No
Transparent data encryption (TDE)	Yes	No
Data discovery & classification	Yes	No
Vulnerability assessment	Yes	No
Advanced threat protection	Yes	No
Auditing	Yes	No
Firewall rules	Yes	Yes
Private endpoint	Yes	Yes

In the Security category, toward the end of the table, you can see that there are many features that are available in Dedicated Synapse SQL Pool, but that are not available in Serverless Synapse SQL Pool. Row-level security (RLS), transparent data encryption (TDE), data discovery & classification, vulnerability assessment, advanced threat protection (ATP), and auditing are some of the important features that are not available in Serverless Synapse SQL Pool. Firewall rules and private endpoints, which are extremely important for network-level security, are present in Dedicated Synapse SQL Pool as well as Serverless Synapse SQL Pool.

There are also some notable differences when it comes to storage-level authentication options with both the pools. While Dedicated Synapse SQL Pool supports storage SAS token authentication, storage access key authentication, and storage application identity authentication, these features are not supported in Serverless Synapse SQL Pool. So, you will have to keep this in mind while deciding on your pool.

Tools

Table 5-4. Tools Feature Comparison

	Dedicated	Serverless
Synapse Studio	Yes, SQL scripts	Yes, SQL scripts
Power BI	Yes	Yes
Azure Analysis Service	Yes	Yes
Azure Data Studio	Yes	Yes, version 1.18.0 or higher. SQL scripts and SQL notebooks are supported.
SQL Server Management Studio	Yes	Yes, version 18.5 or higher

In the Tools category, we have good support available in both options. There are some restrictions with Azure Data Studio and SSMS versions for Serverless option, but apart from that, the other tools are supported in both options equally.

Storage Options

Table 5-5. *Storage Options Feature Comparison*

	Dedicated	Serverless
Internal storage	Yes	No
Azure Data Lake v2	Yes	Yes
Azure Blob Storage	Yes	Yes
Azure SQL (remote)	No	No
Azure CosmosDB transactional storage	No	No
Azure Cosmos DB analytical storage	No	Yes, using Synapse Link (preview) (public preview)
Apache Spark tables (in workspace)	No	PARQUET tables only using metadata synchronization
Apache Spark tables (remote)	No	No
Databricks tables (remote)	No	No

In Storage Options support, there are a couple of notable differences. There is no internal storage for the Serverless option. However, Serverless supports the Azure Cosmos DB analytical store as part of the Synapse Link feature. As Synapse Link is not supported with Dedicated Synapse SQL Pool, the Azure Cosmos DB analytical store is also not supported with it. Similarly, the Serverless option also supports Apache Spark Parquet tables using metadata synchronization, while the same feature is not available in Dedicated Synapse SQL Pools.

Data Formats

Table 5-6. *Data Formats Feature Comparison*

	Dedicated	Serverless
Delimited	Yes	Yes
CSV	Yes (multi-character delimiters not supported)	Yes
Parquet	Yes	Yes, including files with nested types
Hive ORC	Yes	No
Hive RC	Yes	No
JSON	Yes	Yes
Avro	No	No
Delta-lake	No	No
CDM	No	No

The most widely used data formats—like Delimited, CSV, Parquet, and JSON—are supported in Dedicated as well as Serverless Synapse SQL Pool. There are two Hive formats named Hive Optimized Row Columnar and Hive Row Columnar that are supported only in Dedicated Synapse SQL Pool and not in Serverless Synapse SQL Pool. Delta Lake format, which supports creating Delta tables, is not available in either. So, for Delta Lake format, you will have to use Synapse Spark, which we are going to discuss in the next chapter.

Resource Consumption Model for Synapse SQL

The resource consumption model for Synapse SQL is different for both Dedicated Synapse SQL Pool and Serverless Synapse SQL Pool. This is so because each pool uses a different way of provisioning resources within the pool. While a Dedicated Synapse SQL Pool will have the compute resources provisioned specifically for the pool up front before you can start using it, a Serverless Synapse SQL Pool will have the compute resources provisioned on-demand behind the scenes where it is not apparent to users, based on the query workload you submit to it. Due to these differences, the resource consumption models are different for both.

A Serverless Synapse SQL Pool is readily available for consumption upon creation of an Azure Synapse Analytics workspace. Its consumption model does not depend on the compute resources your queries will consume to produce the result, but rather on the size of the data being processed by your queries. As per this resource consumption model, your consumption will be measured and billed per terabyte of data processed by your queries using Serverless Synapse SQL Pool. Now, let us see how the resource consumption model works for Dedicated Synapse SQL Pools.

A Dedicated Synapse SQL Pool uses a DWU- or cDWU-based resource consumption model based on which generation of Dedicated Synapse SQL Pool you are using. Here, DWU stands for data warehouse unit. It is a unit of measure that collectively represents the CPU, I/O, and memory of compute resources. DWU is used to measure the resource consumption for Dedicated Synapse SQL Pool Gen1. Similarly, cDWU, which stands for Compute Data Warehouse Unit, is used to measure the resource consumption for Dedicated Synapse SQL Pool Gen2. It is obvious that to increase the performance of your pool, you will have to increase the DWU/cDWU, and to lower the performance, you will have to decrease the DWU/cDWU. You will be billed based on the DWU/cDWU consumed by you for running your workload—the more the DWU/cDWU, the higher the cost.

Another important concept to understand for the resource consumption model is the Service Level Objective (SLO). Basically, it is a scalability setting that determines the cost as well as performance level of a Dedicated Synapse SQL Pool. The SLO is shown as DW2000 and DW2000c for Gen1 and Gen2 Dedicated Synapse SQL Pools, respectively. Now, Gen2 supports SLO as low as DW100c, which is not the case for Gen1; hence, if you want to use the lower SLOs available in Gen2, you will have to migrate from Gen1 to Gen2.

Synapse SQL Best Practices

We will discuss some of the important Synapse SQL best practices in this section. As Synapse SQL has two different pools—Dedicated Synapse SQL Pool and Serverless Synapse SQL Pool—we are going to discuss the best practices pool-wise. We will start with Best Practices for Serverless Synapse SQL Pool.

Best Practices for Serverless Synapse SQL Pool

- Serverless Synapse SQL Pool, Azure storage account, and Cosmos DB for analytical store components should be provisioned in the same region to minimize latency. By default, the storage account and Serverless Synapse SQL Pool endpoints are provisioned in the same region, but if you plan to connect other storage accounts with Serverless Synapse SQL Pool, then you should ensure that you try to keep it in the same region.

- It is better to use shared access signature (SAS)–based authentication with your storage account than Azure Active Directory (AAD)–based pass-through authentication for better performance.

- It is better to use the Parquet file format for better performance as it is a compressed columnar format. So, you will get better performance than CSV or JSON. It is better to convert CSV and JSON to Parquet format and then read it using Serverless Synapse SQL Pool for improved performance.

- Automatic schema inference at the time of reading the data from files may allow us to write queries quickly and analyze the data stored in those files. However, it comes with its own drawbacks, like degraded performance and incorrect data type inference. So, it is better to check the data types up front and then use that while reading the data for improved performance.

- For better performance, it is better to partition the data appropriately, and later on, at the time of querying the data, it is best practice to specify filename and file path functions so that the query will read and process the data from only the relevant data files. Similarly, while specifying file paths with wild cards, it is better to push the wildcard to the lowest level possible in the file path.

- As a general best practice, those columns that you plan to use in JOIN, DISTINCT, WHERE, ORDER BY, and GROUP BY clauses, you should create the statistics for those columns for better performance. It happens automatically for Parquet files, but for CSV files you will have to create the statistics separately.

- CTAS stands for **C**reate **T**able **As**, which is a very powerful feature if used in the right manner with Serverless Synapse SQL Pool. If you are frequently joining some of the same tables for many of your queries, using CTAS you can create a new table easily and then instead of joining multiple tables within your query, you can use this single table in your queries, which will result in better performance.

Now, let us jump to best practices for Dedicated Synapse SQL Pool in the following section.

Best Practices for Dedicated Synapse SQL Pool

- Dedicated Synapse SQL Pool supports multiple tools for loading and exporting data quickly. It supports Azure Data Factory, PolyBase, and Bulk Copy Program (BCP) for these purposes. However, while loading or exporting a very large volume of data, PolyBase gives the best performance.

- PolyBase is not an ideal solution for querying the data. It must read the entire files when you query them. So, if you are going to use those files in many of your queries, then it is better to load those files once using PolyBase rather than querying them via an external table route.

- Large tables should be distributed using a hash-distribution pattern. We discussed distribution previously in this chapter. It is important to select the correct column for the hash-distribution pattern for optimum performance, unlike in round robin–distributed tables.

- It is best to avoid over-partitioning or under-partitioning as both will impact performance adversely. You need to try to find the optimum partitioning for your given scenario. It is good practice to experiment a little to decide the optimum partitioning strategy for your given data scenario.

- It is also good practice to use dynamic management views (DMVs) to monitor your queries that are getting executed on a Dedicated Synapse SQL Pool. There are many DMVs that provide details about how your queries are getting executed, and if you carefully evaluate those details, you will be able to find many opportunities to optimize your queries.

- A Dedicated Synapse SQL Pool uses resource classes to allocate memory for queries. Larger resource classes are good for improving query performance when you have larger tables, but some queries that may involve pure scans will not give good performance with large resource classes, as they will reduce the concurrency levels. Smaller resource classes are good for increasing concurrency. So, you have to ensure that you use the correct size resource class for the given scenario for optimum performance.

- When you have insert, update, and delete statements in a single transaction, it will take longer to roll them back in an instance of any failures at any stage within that transaction. To avoid longer-running rollback operations, it is best practice to keep the transaction sizes as small as possible.

- It is also good practice to keep your query result size to the minimum possible level. You should look for options to reduce the number of rows getting returned along with the number of columns in it.

- It is also very obvious that if your column size is unnecessarily larger than what your data require then it is going to impact your query performance. So, it is best practice to keep column sizes to the smallest possible size.

How-To's

There are many important activities that you need to know how to perform in order to get the desired end result for Synapse SQL. In the following sections, we are going to discuss how to carry out some of the most important high-level activities step-by-step in Synapse SQL.

Create a Dedicated Synapse SQL Pool

1. Sign in to the Azure Portal and navigate to your Azure Synapse Analytics workspace.

2. You can create a Dedicated Synapse SQL Pool from Azure Portal directly, as well as through Azure Synapse Studio.

    a. To create it using Azure Portal directly, you can see an option named "+ New SQL Pool" on the top menu bar. You need to click on that, which will take you to the wizard through which you can create a Dedicated Synapse SQL Pool.

    b. You also have this option through Azure Synapse Studio. For that, you will have to launch the Synapse Studio from the Azure Synapse workspace. In Synapse Studio, you need to click on the Manage hub link, which is the last link on the left-hand-side navigation. In the Manage hub, click on SQL Pools, and then, on the top menu bar, click the New button to launch the Dedicated Synapse SQL Pool creation wizard.

3. On the Create SQL Pool wizard's Basic tab, you will have to provide the SQL Pool details, which include SQL pool name and performance level, as seen in Figure 5-6.

4. Click on the Next: Additional Settings button at the bottom of the screen. On the Additional Settings tab, you will have to provide data source details and SQL pool collation details, as seen in Figure 5-7.

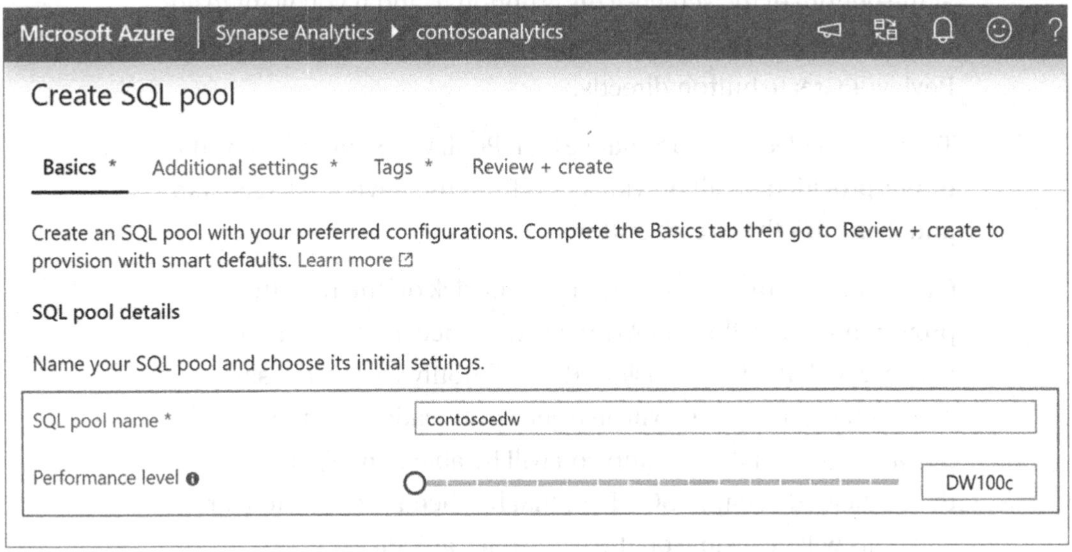

Figure 5-6. Basics tab on Create SQL Pool wizard. Source: Microsoft

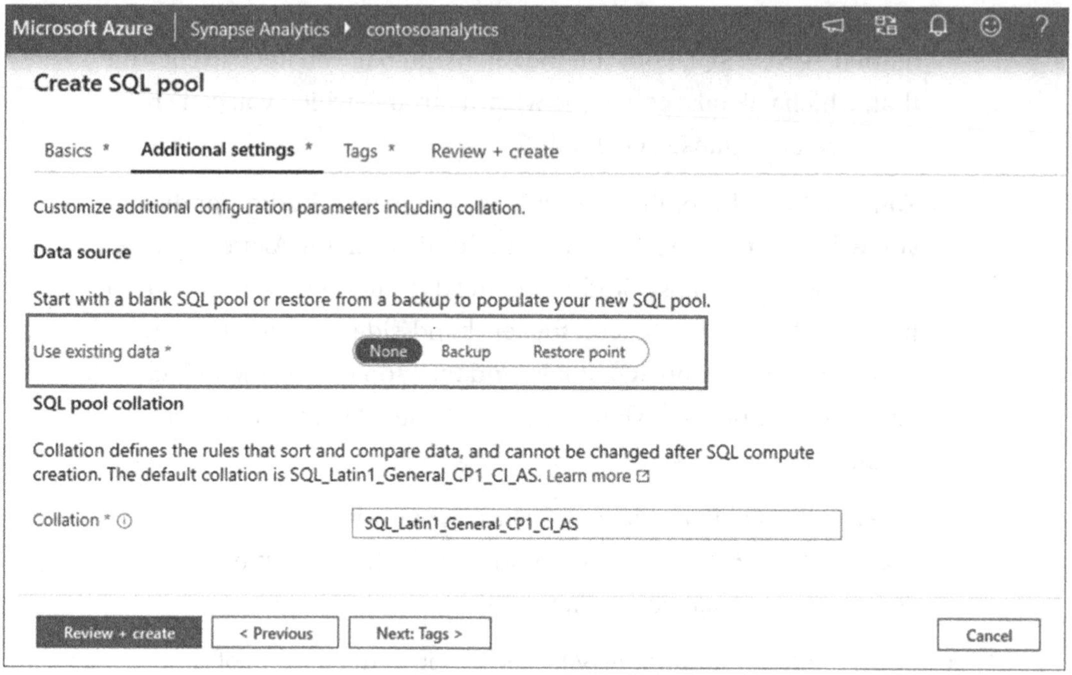

Figure 5-7. *Additional Settings tab on Create SQL Pool wizard. Source: Microsoft*

5. To go to the next tab, named Tags, click on the Next: Tags button at the bottom of the screen. This is optional, and if you want to go directly to the last screen then you can do so by clicking on the Review + Create button directly.

6. To create the Dedicated Synapse SQL Pool, you should review the inputs provided on all previous tabs from the Review + Create tab and then click the Create button.

7. Once you click the Create button, it will kick off the resource provisioning workflow on Azure Portal. Once the Dedicated Synapse SQL Pool is provisioned, it will notify you on the same page. Alternatively, you will also get a notification alert on the portal along with its link, and you will be able to navigate back to the Dedicated Synapse SQL Pool that has been provisioned for you. This will be visible on the workspace as well.

Create a Serverless or On-Demand Synapse SQL Pool

Well, the preceding title is misleading, but intentional. You need not create a Serverless Synapse SQL Pool on your own directly. Actually, when you create and provision a new Azure Synapse Analytics workspace, along with provisioning the workspace, it will also provision an end point for your Serverless Synapse SQL Pool. Once your Synapse workspace is created, you can immediately start using the Serverless Synapse SQL Pool to query the data. This is because it is serverless and so, behind the scenes, Azure provisions this for you while creating the workspace. As there is a way to create a Dedicated Synapse SQL Pool through the portal and Synapse Studio, as we discussed in the previous section, someone may try to find this topic to know how to create a Serverless Synapse SQL Pool.

Load Data Using COPY Statement in Dedicated Synapse SQL Pool

1. The COPY statement is used to bulk load the data into a Dedicated Synapse SQL Pool. As part of the first step, you need to ensure that the right permissions are configured for the user that you are going to use for the COPY statement. Ensure that the user has create table, administer database bulk operations, and insert permissions on the target table.

2. Next, you will execute a CREATE TABLE statement with the following syntax to create the target table in which you want to copy the data:

   ```
   CREATE TABLE [dbo].[TableName] ([Column_1] data_type
   NULL/NOT NULL, [Column_2] data_type NULL/NOT NULL,....,
   [Column_N] data_type NULL/NOT NULL) WITH (DISTRIBUTION =
   distribution_type, CLUSTERED COLUMNSTORE INDEX );
   ```

3. Now, you can run the COPY statement using the following command:

   ```
   COPY INTO [dbo].[TableName] FROM 'Path' WITH (FIELDTERMINATOR=
   '[delimiter]', ROWTERMINATOR='[rowterminator_value]') OPTION
   (LABEL = 'COPY: dbo.[TableName]');
   ```

Ingest Data into Azure Data Lake Storage Gen2

As part of your Azure Data Lake Storage Gen2 in Azure Synapse Analytics, you might like to ingest data from one location to another location within the same Azure Data Lake Storage Gen2 storage account. Let us see how to do this step-by-step.

1. As the first step, we will create a linked service, which will allow us to connect to Azure Data Lake Storage Gen 2 from Synapse Pipelines.

   a. Once you have signed in to Azure Portal, go to your Azure Synapse workspace and launch Synapse Studio.

   b. On the Manage hub of Synapse Studio, under External Connections, you will find a link to create a new linked service.

   c. Select Azure Data Lake Storage Gen 2 and provide related details like authentication credentials and hit the Test Connection button.

   d. If your connection to ADLS Gen 2 is successful, then you should click on the Create button to complete the creation of the linked service.

2. The next step is to create a Synapse Pipelines from Synapse Studio. The pipeline will be using a Copy activity to copy the data from one path of Azure Data Lake Storage Gen2 to another path in it.

   a. You will go to the Orchestrate hub and select the New Pipeline option, which will open a new pipeline canvas on which you can drag and drop the activities you want to include in your pipeline.

   b. Under Move and Transform, you will select Copy Activity and then drag and drop it onto the pipeline canvas.

   c. Select Copy Activity from the pipeline canvas and go to the Source tab, where you need to select New to create a new source dataset for your pipeline. Here, you will select Azure Data Lake Storage Gen2 as your data source and click the Continue button. After that, select Delimited Text as your input format and click the Continue button.

   d. Select the ADLS Gen 2 linked service you created earlier in the Set Properties screen. Then specify the file path of your source data and configure whether the first row has a header or not; click the OK button.

    e. Now, you need to create a new sink dataset. For that, go to the Sink tab and select ADLS Gen2. Again, select Delimited Text as the output format and click the Continue button. Again, select the ADLS Gen2 linked service that you created previously. However, this time you will specify a different folder path in which you want to write your data in ADLS Gen2. Click OK when you are done to complete this activity.

3. As you are done with the pipeline configuration, now it is time to debug and publish your Synapse Pipelines. On the toolbar, click on Debug, and then you should be able to see the status of your pipeline run in the Output tab. Once your pipeline gets executed successfully, click on Publish All, which will publish all the components we have created so far.

4. In the final step, we will manually trigger the pipeline you just published. Click Add Trigger and select the Trigger Now option to trigger the pipeline immediately. Click Finish on the pipeline run screen. For monitoring your pipeline execution, you can go to the Monitor tab located in the left-side navigation bar. You will be able to see the pipeline run that you have just now triggered over there. Wait for your pipeline to finish and then later on you can verify that your data is successfully written at the destination.

Summary

Synapse SQL is an important architectural component of Azure Synapse Analytics. Therefore, we have covered a lot of ground about Synapse SQL and covered many important topics about it in this chapter. We started with an overview of Synapse SQL architecture components and discussed Dedicated Synapse SQL Pool as well as Serverless Synapse SQL Pool in detail. We covered massively parallel processing (MPP) and distributed query processing (DQP) engines, as Dedicated and Serverless Synapse SQL use these engines, respectively.

Both MPP and DQP are node-based architectures and consist of a Control node and one or more Compute nodes. The Control node is the entry point for any application or user, and the actual computation happens in the Compute nodes. Data movement service, which is present in Dedicated Synapse SQL only, is responsible for moving the data across the nodes as and when required. We also looked at how data gets distributed in storage and various table distribution patterns like round robin, hash-distributed tables, and replicated tables.

We discussed high-level details about Dedicated Synapse SQL Pool followed by Serverless Synapse SQL Pool. They are different architectures and provide slightly different capabilities. We also looked at and compared some of the important feature groups, like database object types, query languages, security, tools, storage options, and data formats. The comparison helped us to understand some of the differences between the two pools that play an important role when deciding which pool to use for which use case.

It is also very important to understand the resource consumption models for both pools. We discussed the Dedicated Synapse SQL Pool resource consumption model followed by the Serverless Synapse SQL Pool resource consumption model. While Dedicated Synapse SQL Pool uses a DWU/cDWU–based resource consumption model, Serverless Synapse SQL Pool uses a resource consumption model based on each terabyte of data processed. There are some best practices while using both the pools, which we discussed at a high level in this chapter. Those best practices will really be handy while troubleshooting any technical issues for Synapse SQL.

It is important that you are aware of how to carry out certain important activities related to Synapse SQL. Therefore, we looked at some important how-to's with step-by-step processes to carry out those activities successfully. As part of that section, we covered how to create a Dedicated Synapse SQL Pool, how to create a Serverless Synapse SQL Pool, how to use the COPY statement to load data using a Dedicated Synapse SQL Pool, and how to ingest data into Azure Data Lake Storage Gen2.

With all the preceding topics, we have discussed Synapse SQL and its core components in detail, with some of the comparisons highlighting differences between Dedicated and Serverless Synapse SQL Pools. Another important component—and the most exciting one as well—is Synapse Spark, which is Microsoft's implementation of Apache Spark for Azure Synapse Analytics. We are going to discuss this in more detail in the next chapter. So, stay tuned and see you there!

CHAPTER 6

Synapse Spark

Azure Synapse Analytics provides three compute engine options to choose from. There are two Synapse SQL engines, namely Serverless Synapse SQL and Provisioned Synapse SQL; the third option is Synapse Spark, which is based on Apache Spark. We discussed Synapse SQL in detail in the previous chapter. Now, let us discuss Apache Spark, or Synapse Spark, in detail in this chapter.

Big Data analytics requires a robust data analytics engine that can support petabyte-size processing with fast processing power. Apache Spark fits the bill perfectly. In this chapter, we will look at Synapse Spark in detail. We will start our discussion with what Apache Spark is and what Synapse Spark is within Azure Synapse Analytics. Synapse Spark provides multiple features and capabilities for Azure Synapse Analytics. It is one of the most important additions to Azure Synapse Analytics on top of the legacy Azure SQL Data Warehouse. We will look at some of its important features and capabilities before moving to another important topic—Delta Lake. The combination of Apache Spark and Delta Lake is a very good one for many data analytics scenarios. We will discuss the importance of Delta Lake in the Spark ecosystem and what benefits it brings to the table if we use it with Synapse Spark.

Spark is a complex platform and is a little difficult to optimize for a specific data analytics workload. Hence, we are going to look at some of the important aspects of Synapse Spark job optimization. These optimization techniques will help you to understand the process of fine-tuning Synapse Spark's performance. Synapse Spark job optimization techniques will allow you to save money and time while running your jobs on a Synapse Spark Pool.

Synapse Spark is a good option for your artificial intelligence workload as well. It can be used for machine learning purposes very easily as it has built-in support for it. We will discuss in detail how you can utilize a Synapse Spark Pool to run your various machine learning workloads. Toward the end of the chapter, we are going to look at some of the important activities for Synapse Spark and how you can perform them. We will look at those activities step-by-step to make it easy to understand and comprehend. So, let us get started!

© Bhadresh Shiyal 2021
B. Shiyal, *Beginning Azure Synapse Analytics*, https://doi.org/10.1007/978-1-4842-7061-5_6

What Is Apache Spark?

Apache Spark is an open source cluster-computing technology that provides a lightning fast unified analytics engine that can easily cater to very large-scale data processing. It is a parallel processing framework that is capable of in-memory processing to provide the fastest possible performance for Big Data analytics applications. Compared to legacy Hadoop systems, Apache Spark can run data processing workloads almost 100 times faster. Currently, Azure Synapse Analytics supports Apache Spark 2.4. You can visit https://spark.apache.org/ for more information on Apache Spark.

Apache Spark provides the following high-level APIs, using which you can write your code for Apache Spark:

1. Java API

2. Scala API

3. Python API (pySpark)

4. R

Apache Spark also supports the following set of high-level tools:

1. Spark SQL

2. MLlib for machine learning,

3. GraphX for graph processing

4. Structured Streaming for incremental computation and stream processing

Prior to Apache Spark, Hadoop was widely used, and even today there are a plethora of organizations using Hadoop. But slowly and steadily, organizations are moving away from legacy Hadoop-based systems and either migrating to Apache Spark–based systems or transforming their applications and then moving to Apache Spark to utilize the real potential of the Apache Spark engine. Let us try to understand how Apache Spark processes the data differently than Hadoop. Please study Figure 6-1 very carefully.

Hadoop uses the traditional MapReduce method to apply transformations on data and carry out data processing. You can easily see that it has to read the data to be processed from the Hadoop Distributed File System (HDFS) and then apply the transformations on that data, and later it must write the transformed data back to the HDFS. That means that for each step within your Hadoop-based traditional MapReduce job, you have to write the data back to persistent disk storage before performing the next step on the same data. This impacts the data processing performance adversely. Compared to that, in the case of Apache Spark, the data processing happens in an entirely different manner.

The Apache Spark engine reads and writes data to the HDFS, but it can carry out all the steps in parallel and in-memory without writing the intermediate data back into the HDFS. Once all the transformation steps are complete, toward the end only, Apache Spark writes the data back to the HDFS. This way, the data processing speed increases drastically compared to what you get in traditional MapReduce-based Hadoop jobs. Parallel in-memory processing is the primary reason Apache Spark provides far better performance than Hadoop.

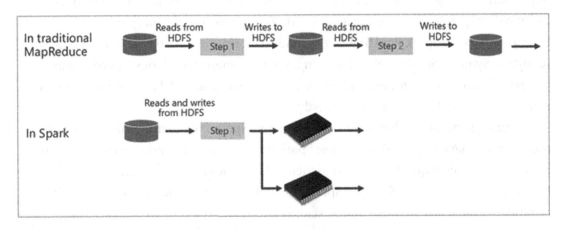

Figure 6-1. *Apache Spark vs. Hadoop MapReduce. Source: This image is taken from Microsoft Documentation titled "Apache Spark in Azure Synapse Analytics" under the sub-section titled "What is Apache Spark" dated April 15, 2020.* `https://docs.microsoft.com/en-us/azure/synapse-analytics/spark/ apache-spark-overview#what-is-apache-spark`

Apache Spark supports a wide array of different workload types easily and inherently. You can use Apache Spark to write batch applications that can be executed on a time-based schedule. It also supports interactive queries, which makes developing code a lot easier. Apache Spark allows you to develop your machine learning algorithm iteratively. As mentioned earlier, Apache Spark provides Structured Streaming also, which can be utilized to implement real-time or near-real-time integration scenarios.

What Is Synapse Spark in Azure Synapse Analytics?

Apache Spark is an open source project. It is driven by a very large set of community members. Microsoft already provides a Spark engine as part of two different offerings on Azure. The first is through Azure HDInsight, and the second is through Azure Databricks. Both these services use Apache Spark as the foundation. Along with Azure Synapse Analytics, Microsoft introduced its own implementation of Apache Spark. Therefore, you additionally get support for C#.Net to be used as one of the programing languages for Synapse Spark.

Generally, it is referred to as Apache Spark or Spark, but in order to differentiate spark in Azure Synapse Analytics from the general Apache Spark or Spark implementation, we are going to refer Microsoft's implementation of Apache Spark in Azure Synapse Analytics as Synapse Spark. The creation and configuration of a Serverless Apache Spark Pool in Azure is made a lot simpler and easier in Azure Synapse Analytics. Synapse Spark Pools created in an Azure Synapse Analytics workspace are compatible with Azure storage and Azure Data Lake Storage (ADLS) Gen2 so that you can process the data stored in Azure easily.

In Azure Synapse Analytics, you use a Synapse Spark Pool to utilize the power of the Synapse Spark engine. The Synapse Spark Pool is a set of metadata that defines the compute resource requirements and associated behavior characteristics when a Spark instance is instantiated. A Synapse Spark Pool defines many characteristics like its name, number of nodes, size of those nodes, scaling configuration, time to live, etc. These characteristics will decide how your Synapse Spark Pool is going to behave when you execute some workload on it. There is no charge to define and keep a Synapse Spark Pool in your Azure Synapse Analytics workspace. You will only be charged based on the usage of your Synapse Spark Pool, and that usage will be decided by the type of workload, which will determine the number and size of Worker nodes to be used.

The Synapse Spark Pool will consist of one Header node and two or more Worker nodes. That means that there will be a minimum of three nodes in any Synapse Spark Pool. All Worker nodes will be running the Spark Executor service, while the Header node will be running the Apache Livy, Yarn Resource Manager, Zookeeper, and Spark Driver services. Both the Header node and Worker nodes will be running Node agent and Yarn Node Manager. You have multiple options when you select node size. Node size will determine the number of vCores and memory available to you from a node. You can select a small node with 4 vCore and 32 GB of memory, and on the higher side you can select a node having 64 vCore and 432 GB of memory.

Synapse Spark Features & Capabilities

Synapse Spark provides many features and capabilities in Azure Synapse Analytics. It is important to understand these features and capabilities so that you can extract the maximum benefit out of Synapse Spark, and this will also help you to understand what is supported in Synapse Spark that can be used in your project directly. Let us discuss those features and capabilities one by one.

Speed

Synapse Spark runs data analytics workloads 100 times faster than legacy Hadoop–based MapReduce jobs. Internally, Synapse Spark uses a state-of-the-art directed acyclic graph (DAG) scheduler, a query optimizer, and a physical execution engine to provide high performance for both batch and streaming data. Due to its data processing speed, Apache Spark has become a de facto standard for any large-scale data processing workloads. That same engine is available to you in the form of Synapse Spark in Azure Synapse Analytics.

Faster Start Time

In Azure Synapse Analytics, Synapse Spark instances will start in approximately two minutes if you have fewer than 60 nodes in your cluster. If you have more than 60 nodes in your cluster, then it will take approximately five minutes to start the instances. The Synapse Spark instance will shut down, by default, approximately five minutes after the last job executes. If you want to keep the instance alive, then you need to ensure that you keep a notebook connection to the instance; that way, it will not get itself shut down as it will have some work to do from your notebook and will not be sitting idle, which may result in auto-shutdown.

Ease of Creation

It is very easy to create a Synapse Spark Pool in Azure Synapse Analytics. If you prefer to use a GUI-based option, then you can create a new Synapse Spark Pool in Azure Synapse Analytics in minutes using the Azure Portal. Alternatively, you can use Azure PowerShell or Synapse Analytics .NET SDK to create a Synapse Spark Pool, which allows you to automate the process of Synapse Spark Pool creation using either of the two options.

Ease of Use

As part of Synapse Studio within the Azure Synapse Analytics workspace, you get a very easy-use option to write your notebooks. You get an option to write your notebooks using a customized experience that is derived from Nteract, which is an open source interactive computing environment project. These notebooks provide an easy way to run your code and visualize your data using Synapse Spark Pool. This notebook experience is similar to other notebook writing experiences in which you get in Azure Databricks with some minor differences.

Security

Spark supports multiple levels of security, enabled by you, based on your deployment type. In our case, Azure Synapse Analytics deploys Spark on YARN. So, it handles the generation and distribution of shared secrets automatically. This is required for making a secured connection to Spark. Currently, Spark supports authentication for RPC channels using a shared secret. It also supports AES–based encryption for RPC connections. Additionally, Spark also supports encrypting temporary data written to local disks, which include shuffle files, shuffle spills, and data blocks stored for broadcast and cache purposes. Spark also supports the SSL configuration for secure communication.

Automatic Scalability

As we know, Apache Spark is a node-based architecture in which you have a Header node and two or more Worker nodes. Synapse Spark allows you to set up automatic scalability for Worker nodes in the Synapse Spark Pool. It is very easy to configure auto-scale, as you define the minimum and maximum numbers of Worker nodes that you expect your workloads to use. During runtime, based on the workload you provide to your Synapse Spark Pool, it will either increase or decrease the number of Worker nodes your workload requires to perform the tasks on hand. Another feature to note here is that you also have the option to set up an automatic shutdown of interactive clusters so you can save on compute cost as well.

Separation of Concerns

Synapse Spark provides a compute engine that can process your data at a rapid speed. In Azure Synapse Analytics, you can easily store your data in either an Azure Data Lake Storage (ADLS) Gen2 or Azure Blob Storage account, as both are available for direct integration with Azure Synapse Analytics. The data stored in either ADLS Gen2 or Azure Blob Storage can easily be used by Synapse Spark Pool. Data is stored separately from the compute engine, so it separates our concerns around storage and compute environments, as they are independent of each other. Whenever not required, you can bring down your Synapse Spark Pool without worrying about the loss of any data, as your storage can exist independent of your compute environment, which is Synapse Spark.

Multiple Language Support

Synapse Spark provides the option to code in different programming languages using its Nteract-based notebook experience. It supports the following languages for this purpose:

1. Python (pySpark)

2. Spark (Scala)

3. .NET for Spark

4. Spark SQL

It is important to note that Synapse Spark does not support R or Java, which are supported by Apache Spark. However, Microsoft has introduced .NET for Spark, which allows C# developers to code for Synapse Spark without a steep learning curve, which is obvious for those who have to learn to code in either pySpark or Scala. Microsoft has open-sourced the .NET for Spark libraries, which means that you can further customize this library as per your requirements or take advantage of customizations done by open source community contributors.

Integration with IDEs

There are situations in which you are not allowed to code using the out-of-box Nteract-based notebook development experience. In that situation, Synapse Spark allows you to connect a third-party integrated development environment (IDE). Azure Synapse

Analytics provides an IDE plug-in that you can use with JetBrains' IntelliJ IDEA. This plug-in can help you to create and submit applications for execution to Synapse Spark Pool. This allows you to develop your code outside the Azure Synapse Analytics workspace and execute that code directly on the Synapse Spark Pool. There is also an option to connect Visual Studio Code, which is an open source IDE from Microsoft, with Synapse Spark Pool.

Pre-loaded Libraries

In Azure Synapse Analytics, the Synapse Spark Pool comes pre-loaded with many libraries you can use within your code and run on the Synapse Spark Pool. Anaconda is a freely available distribution of over 7,500 open source packages. It is the most trusted distribution for data science. Anaconda is pre-installed in the Synapse Spark Pool for you. That means that you can utilize all those packages and libraries for machine learning, data analysis, and data visualization without needing to install them individually on your Synapse Spark Pool. It makes developers' work easy, as most of the widely used libraries are already pre-installed in the Synapse Spark Pool.

REST APIs

Synapse Spark includes Apache Livy, which is an open source project. It is a REST API–based Spark job server that allows you to submit the jobs to Spark and monitor them remotely. As it is by default available with Synapse Spark Pools as well, you can remotely submit your application to Synapse Spark and monitor it remotely as well. This adds flexibility to manage everything remotely without visiting the Synapse Spark Pool directly. It allows multiple users to interact with your Synapse Spark Pool concurrently and reliably. For more information on this, you can visit `http://livy.incubator.apache.org`.

We have seen many features and capabilities of Synapse Spark in the preceding sections. Now, let us move ahead and look at what Delta Lake is and what its importance is in the Synapse Spark world.

Delta Lake and Its Importance in Synapse Spark

We have discussed Delta Lake briefly at a high level in previous chapters. Let us recap some of this. It is an open source storage layer that brings ACID (Atomicity, Consistency, Integrity, and Durability) transactions to Synapse Spark and Big Data workloads that you can execute using Synapse Spark. Delta Lake brings reliability to your data lake, which is Azure Data Lake Storage (ADLS) Gen2 in our case. Delta Lake runs on top of your existing data lake and is fully compatible with Synapse Spark as well.

Support for ACID transactions is really very crucial for your Synapse Spark Pool for various reasons. Generally, you populate your data lake via multiple different data pipelines and processes. Many times, you will face a situation in which you are writing and reading the data to your data lake concurrently. Without Delta Lake, it is very difficult to maintain data integrity, because data engineers will have to go through a manual and error-prone method to maintain the data integrity in your data lake. With Delta Lake, we have transaction logs, which help support ACID transactions within your data lake. ACID transactions in Delta Lake are not as good as what you get in a typical Relational Database Management System (RDBMS), but Delta Lake tries to make it as close as possible given the nature of data storage. Without Delta Lake, it is not possible to achieve ACID transactions in your data lake so easily. Hence, it is necessary to have Delta Lake implemented on top of Azure Data Lake Storage (ADLS) Gen2 for your Synapse Spark Pool. Delta Lake and Synapse Spark are tightly integrated. You can easily process the data stored in Delta Lake table format in your data lake using Synapse Spark Pool.

Apart from ACID transactions, which is one of the most important features, Delta Lake brings many other features and capabilities to Synapse Spark Pools. Let us look at them at a very high level here:

- Delta Lake relies on Apache Parquet format, which is again an open source format. It brings efficient compression and encoding schemes.

- Delta Lake also provides unified batch and streaming source and sink. It means that a table stored in Delta Lake is both a batch table and a streaming source and sink.

- Delta Lake brings time travel for your data changes through data versioning, which makes it possible to query different versions of the same data using Synapse Spark.

- It also supports schema enforcement, which means that it can check if the data types are correct and whether all the required fields are present or not. This helps in avoiding bad data that results in data inconsistencies.

- If you want to delete, update, insert, or merge your data using a Synapse Spark Pool then do not worry as Delta Lake supports all those operations.

There are many features and capabilities in Delta Lake that are very useful for your Synapse Spark Pool. Without Delta Lake in place on top of your Azure Data Lake Storage (ADLS) Gen2, you will not be able to get the benefits of all these features and capabilities in your Synapse Spark Pool. Hence, it is extremely important that you implement Delta Lake in Azure Synapse Analytics, which will allow you to explore the full potential and all the capabilities that Synapse Spark brings to the table as a unified, lightning fast data analytics platform.

Synapse Spark Job Optimization

Synapse Spark jobs are submitted to the Synapse Spark Pool for execution. It is necessary that the submitted jobs give optimum performance so as to meet the service-level agreements (SLAs) defined in your business requirements. The performance of a Synapse Spark job depends on multiple aspects, like memory management, caching, Spark cluster configurations, and more. Let us discuss some of the common Synapse Spark job optimizations you need to take into consideration to ensure the best performance.

Data Format

Synapse Spark supports several different data formats for your workloads. You can use the following data formats with Synapse Spark:

- CSV: Comma-separated values

- JSON: JavaScript Object Notation

- XML: eXtensible Markup Language

- Parquet: Open source columnar data format from Apache

- AVRO: Row-oriented remote procedure call and data serialization

- ORC: Optimized Row Columnar file format

Apache Parquet is the best data format, and that is why it is used widely in large-scale data analytics workloads. It is open source and free to use, and supports many data processing frameworks, data models, and programming languages. Parquet with Snappy compression is the most performant combination. You also have the option to use gzip compression along with Parquet. In order to get the best performance from your Synapse Spark jobs, you need to evaluate all these data formats and select the most appropriate one for your given situation. For example, large-sized CSV will always be time consuming to read compared to Parquet, but if you have very small file sizes then CSV may work fine.

Memory Management

Spark uses in-memory data processing, so it is very important that you manage the available memory efficiently to get the best performance from your Synapse Spark jobs. Internally, Synapse Spark uses YARN (**Y**et **A**nother **R**esource **N**egotiator), which is an open source and free-to-use resource manager available from the Apache Hadoop YARN project. YARN is responsible for controlling the maximum sum of memory used by all the executors that will be running in each Synapse Spark node. "Out of Memory" is a common error or issue you will face while running your jobs on a Synapse Spark Pool. However, by following some of the best practices around memory management you should be able to overcome this issue easily. It is recommended that you use DataFrame API instead of low-level RDD API. You should also try to avoid data shuffling, which may happen across nodes. The partitioning of your data also plays an important role, so ensure that you partition your data appropriately so that you don't have large variations in partition sizes.

Data Serialization

Data serialization is the process of converting objects into a stream of bytes and vice versa. It plays an important and crucial role in any distributed application, and hence it is important for us to handle data serialization in the most efficient manner. Synapse Spark provides the following two data serialization options to choose from:

- Java: This is the default serialization available in Synapse Spark. It is less efficient than Kryo serialization.

- Kryo: This is a relatively newer format that can give you faster and more compact serialization than what you get from Java serialization. You will have to enable Kryo serialization by making a configuration change in your Synapse Spark environment.

Data Caching

If you are going to use the same data very frequently from your storage, then it makes sense to store that data in the cache to get the best performance possible from your Synapse Spark job. Spark provides three different caching mechanisms, as follows:

- `.persist()`: Stores in user-defined storage level

- `.cache()`: By default saves to memory only

- `CACHE TABLE`: Eager operation to cache whole table to storage, which gets executed along with the statement, as it is not a lazy operation

These native caching mechanisms are useful for smaller datasets as well as for storing intermediate results from data pipelines. Synapse Spark's native cache mechanisms do not work well with partitioned data. Based on the requirements of your Synapse Spark workload, you need to decide which caching mechanism is most effective from a performance perspective. The caching of data to memory or storage will itself have some operational performance overhead, so keep that in mind while deciding your caching strategy.

Data Abstraction

Synapse Spark provides three different APIs for data abstraction: RDD, DataFrame, and DataSet. RDD stands for Resilient Distributed Datasets. RDD is the low-level API and is a legacy option for data abstraction. DataFrame and DataSet got introduced to Spark after RDD. Since RDD is a low-level legacy API, generally it is not widely used nowadays.

It does not provide you any query optimization through the Catalyst engine. There is also an additional high level of GC (garbage collection) overhead. Therefore, it is best to avoid using RDD.

DataFrame is widely used as it suits most scenarios. It uses Catalyst to optimize queries for better performance. It has a low level of GC overhead compared to RDD. DataFrame also supports direct memory access and whole-stage code generation in Synapse Spark. However, it has a drawback also. DataFrame is not considered developer-friendly as it is not possible to carry out any compile-time checks, so your code may show errors during execution. DataFrame does not support domain object programming.

Compared to RDD and DataFrame, DataSet is considered developer-friendly as it supports compile-time checks as well as domain object programming, which are not supported features in RDD and DataFrame. DataSet is good for complex ETL pipelines where performance impact is acceptable but not good for aggregation operations where performance impact can be very considerable. Thus, to get the optimum performance from Spark's data abstraction concept, you need to use the appropriate API based on the type and complexity of your data analytics workload.

Join and Shuffle Optimization

Generally, joins are unavoidable in data analytics workloads. By default, Synapse Spark supports the `SortMerge` type of join. In a `SortMerge` join, Synapse Spark sorts the left and right sides of data and then merges them together. This type of join is required for large sets of data, which will impact the performance of your Synapse Spark job adversely. However, if you are joining one large dataset with another small dataset, then `SortMerge` can be avoided easily. In this scenario, you have the option to use the `broadcast` type of join, which is more efficient and will give better performance for the given scenario. Ideally, you should broadcast the smaller table to all the Worker nodes so that each node will be able to perform the join with the larger table without shuffling the data across nodes. This saves a lot of unnecessary data movement and will result in cost and time savings as well.

The order of joins also plays a crucial part in your query performance. For example, the most selective join should be applied first so that it will filter out all the unnecessary rows up front so that you do not have to apply any complex aggregations on those rows that are going to get filtered or removed later in the query execution anyway.

Bucketing

Partitioning and bucketing are similar, but there is a slight difference. Generally, in a partition, you have all those rows that belong to just one partition key. However, in bucketing, you have an option to have a set of column values rather than just one. So, if you have a very large set of unique values, then it is most appropriate to implement it. By hashing the bucket key of the row, a bucket is determined internally by Synapse Spark. Bucketing offers many advantages, including optimized joins and aggregations. Bucketing and partitioning can be used together to optimize the performance of your Synapse Spark jobs.

Hyperspace Indexing

Out of the box, Apache Spark and therefore Synapse Spark as well does not support the indexing of data. However, Hyperspace allows you to index the data for Spark. Hyperspace is an early-phase indexing subsystem for Apache Spark that introduces the ability for users to build indexes on their data, maintain them through a multi-user concurrency mode, and leverage them automatically without any changes to their application code for query or workload performance optimization.

It is very simple to enable Hyperspace for your Synapse Spark job as it does not require any big changes in your existing codebase. Similarly, if you do not want to use Hyperspace indexes after you have created them, you can disable them without any big changes to your existing codebase.

Synapse Spark Machine Learning

Synapse Spark supports many machine learning capabiltiies within Azure Synapse Analytics. The data science process contains fairly common steps in this specific sequence: data acquisition, data preparation, modeling, model deployment, and model scoring. Many of these steps can easily be carried out by using Synapse Spark.

Data Preparation and Exploration

Synapse Spark can help you to prepare, transform, and explore data at scale easily. You can create Synapse Spark notebooks to run your code in multiple supported langaguges. You can use PySpark (Python), Scala, and .NET for Spark for this purpose. During data preparation or data exploration, you often need to visualize the data so that you understand it better and can generate quicker insights. Synapse Spark supports many different visualization libraries that can be used to visualize the data with Synapse Spark.

Build Machine Learning Models

Synapse Spark can utilize Apache Spark's MLLib to build machine learning models. With MLLib, you can solve most of the classical machine learning problems by using appropriate libraries and algorithms to solve a specific given problem. You also have the option to use Sci-kit learn libraries to develop your machine learning models using Synapse Spark. Due to the iterative capability of Nteract-based notebooks, a data science team can quickly build and run the models using Synapse Spark from Synapse Studio. That makes it easy and quick to develop new machine learning models with Synapse Spark.

Train Machine Learning Models

Spark's MLLib and Sci-kit learn libraries can also be used to train your machine learning models with your data at scale using Synapse Spark. There are many other libraries as well that can easily be installed on Synapse Spark and can be used for training your machine learning models. Apart from this, you also have the option to use Automated ML from Azure Machine Learning for training your machine learning models. Automated ML, or AutoML, is a feature that automatically trains a set of machine learning models and allows the user to select the best model based on specific metrics.

Synapse Studio provides seamless integration with AutoML and Synapse Spark. This integration allows you to develop the notebook using Synapse Studio, in which you can utilize AutoML from Azure Machine Learning. Due to AutoML's leveraging passthrough authentication using Azure Active Directory, the users of notebooks in Azure Synapse Analytics can easily connect AutoML to train their machine learning models automatically. This makes it easy to train multiple different models using Synapse Spark and compare the model metrics to select the most appropriate one.

Model Deployment and Scoring

Azure Synapse Analytics provides the option for batch scoring your machine learning models. For this purpose, you can leverage a Synapse Spark Pool to deploy your machine learning model and score it to understand and fine-tune the model behavior so as to meet business requirements. Depending on the libraries used to train your machine learning models, you can use a code experience on your batch scoring, which can be run using Synapse Spark.

How-To's

We have gone through some of the important components, features, and capabilities of Synapse Spark. However, we have not looked at how to carry out a specific activity or task with it. So, now we will discuss how you can perform a few of the important tasks for or using Synapse Spark.

How to Create a Synapse Spark Pool

1. Log in to your Azure subscription by opening the Azure Portal and open the Azure Synapse Analytics workspace under which you want to create a Synapse Spark Pool.

2. Within your Azure Synapse Analytics workspace's Overview section, you will see an option to launch Synapse Studio. Please launch Synapse Studio using that option.

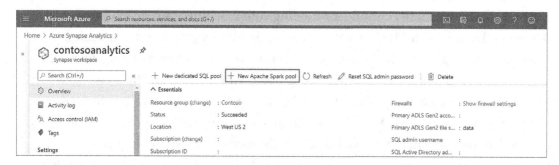

Figure 6-2. *New Apache Spark Pool option. Source: Microsoft*

3. Click on the + New Apache Spark Pool button to launch the wizard for creating an Apache Spark Pool, as shown in Figure 6-2.

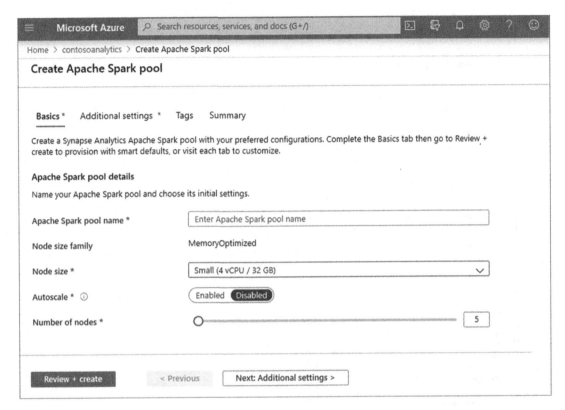

Figure 6-3. *Basic tab on Create Apache Spark Pool wizard. Source: Microsoft*

4. Enter the following details on the Basic tab of the wizard, as shown in Figure 6-3:

- **Apache Spark Pool Name:** Provide a valid name for the Apache Spark Pool you want to create.

- **Node Size:** Select the node size from the dropdown as per your requirements.

- **Autoscale:** You can either enable or disable the autoscale option for your Spark Pool.

- **Number of Nodes:** Select the number of nodes you want to have in your new Spark Pool.

5. Click on the Next: Additional Settings button to go to the next
    screen within the wizard, as shown in Figure 6-4.

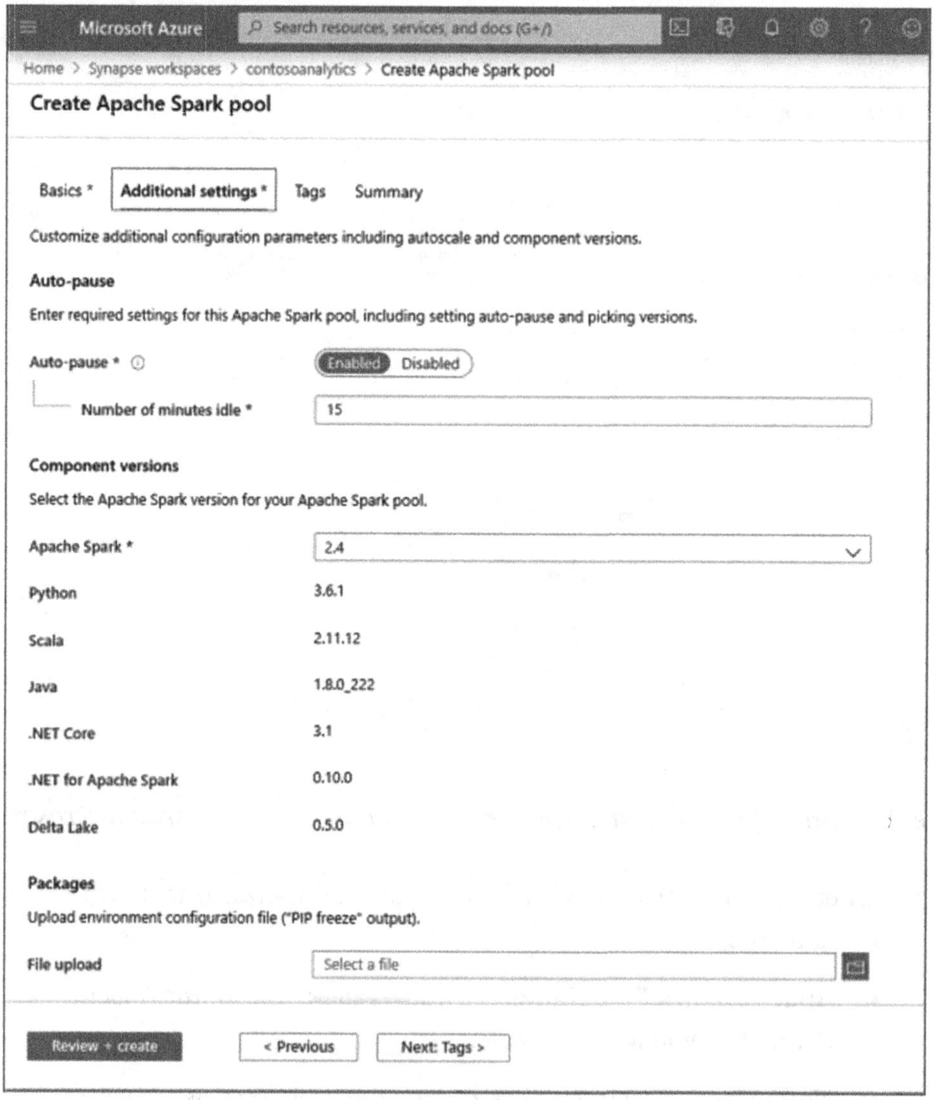

Figure 6-4. *Additional settings for creating Apache Spark Pool. Source: Microsoft*

You will have options to modify the following settings or keep them as the default without making any changes:

- **Auto-pause:** You can either enable or disable this setting, which will decide if the Spark Pool needs to be automatically paused after a certain number of minutes or not.

- **Number of Minutes Idle:** You can enter the number of minutes the Spark Pool should remain idle before automatically pausing it. If you keep the auto-pause setting in a disabled state, then this will not have any impact.

- **Component Versions:** Here, apart from the version for Apache Spark, everything else is auto-selected for you. These are read-only values and will depend on what value you select for Apache Spark version.

- **File Upload:** It allows you to select a file containing the environment configuration that will be used for your Spark Pool.

6. Click on the Next: Tags button, which will take you to the Tags screen, as shown in Figure 6-5, on which you can define tags with their names and values as per your requirements.

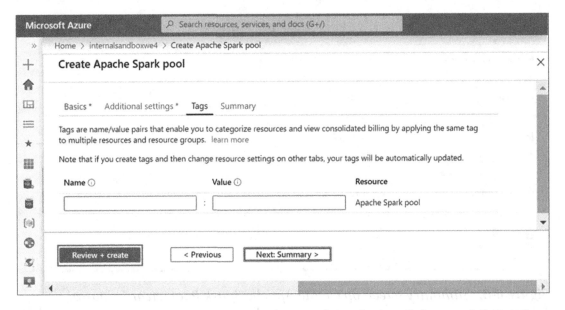

Figure 6-5. Tags screen on Create Apache Spark Pool wizard. Source: Microsoft

You can create tags as name–value pairs or leave this screen without any changes if you do not want to have any tags for your Spark Pool.

7. Click on the Review + Create button, which will take you to the summary screen, as shown in Figure 6-6, which will display all the settings you have configured on the previous screens of the Create Apache Spark Pool wizard.

Figure 6-6. *Summary screen on Create Apache Spark Pool wizard. Source: Microsoft*

Verify all the details from this screen and ensure that everything is correct as per your requirements, and then click the Create button to kick off the resource provisioning workflow.

8. Once your Apache Spark Pool is created in your Azure Synapse Analytics workspace, it will show you the "deployment is complete" status, as shown in Figure 6-7.

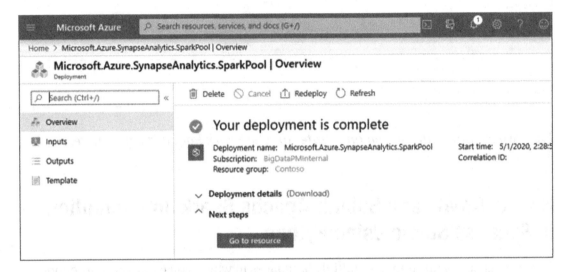

Figure 6-7. *Summary screen on Create Apache Spark Pool wizard. Source: Microsoft*

9. After the provisioning is complete, you can navigate to your Azure Synapse Analytics workspace, where you will be able to see the newly created Apache Spark Pool, as shown in Figure 6-8.

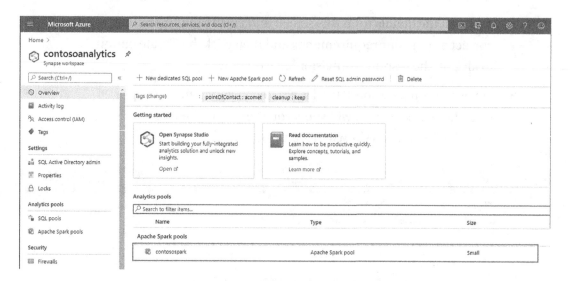

Figure 6-8. *Summary screen on Create Apache Spark Pool wizard. Source: Microsoft*

How to Create and Submit Apache Spark Job Definition in Synapse Studio Using Python

Azure Synapse Studio supports multiple languages in which you can write a notebook that can be submitted to an Apache Spark Pool as a job definition for execution. Let us look at how to create an Apache Spark job definition in Synapse Studio using Python 3.6 or PySpark. PySpark documentation is available at `https://spark.apache.org/docs/latest/api/python/index.html`. Once you know how to do it using one language, you should be able to do the steps for other languages as well.

1. Log in to your Azure subscription by opening the Azure Portal, then open the Azure Synapse Analytics workspace and launch Synapse Studio from it.

2. You should have your Python code file and data file ready so that you can upload them to Azure Data Lake Storage Gen2 using Synapse Studio. If you do not have your own Python code file and data file with you, then you can download a sample Python code file and a data file from the following location: `https://github.com/Azure-Samples/Synapse/tree/master/Spark/Python`. Here, we will assume that you are using these sample files to create an Apache Spark job definition in Synapse Studio.

3. Unzip the downloaded compressed package and extract both files, as shown in Figure 6-9.

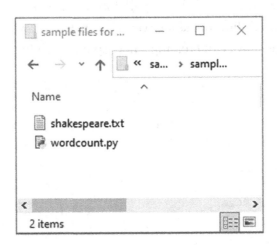

Figure 6-9. *Sample files from GitHub. Source: Microsoft*

4. In Synapse Studio, click on the Data hub and then click on the Linked tab. It will show you the Azure Data Lake Storage Gen2 account. Upload both files into the ADLS Gen2 file system as per Figure 6-10.

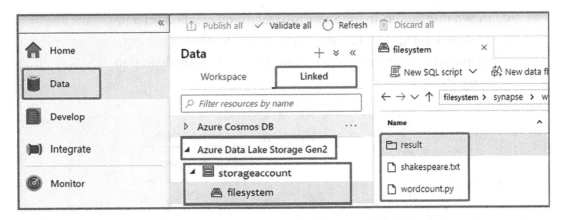

Figure 6-10. *Upload files to ADLS Gen2 using Synapse Studio. Source: Microsoft*

5. In Synapse Studio, click on the Develop hub and then click on
    the "+" icon, which will open a menu containing many options
    to create new artifacts. Click on Spark Job Definition to create a
    new Spark Job Definition, as shown in Figure 6-11. It will open a
    development canvas for creating a new Spark job definition.

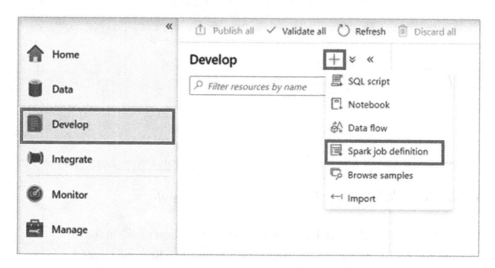

Figure 6-11. *Create new Spark job definition from Develop hub in Synapse Studio.*
Source: Microsoft

6. In Synapse Studio, in the Spark Job Definition creation screen,
    as shown in Figure 6-12, select PySpark(Python) from the top
    dropdown, as we are going to create a new Spark job definition
    using Python.

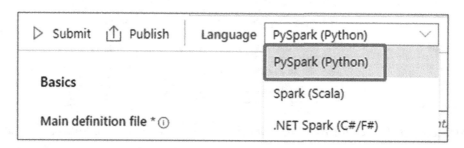

Figure 6-12. *Select PySpark (Python) as language for Apache Spark job definition.*
Source: Microsoft

7. On the Spark Job Definition main window, there are two different sections where you have to fill in details pertaining to the Spark job definition you want to create using Python in Synapse Studio. These two sections are Basics and Submission Details, as shown in Figure 6-13.

- In the Basics section, you will have to fill in the following details:

Job Definition Name:

You should enter an appropriate name for your Apache Spark job definition. When you publish your job, this name will be used, so you can change the name before publishing the job but not after it is published.

Main Definition File:

You should select the main Python file that you want to execute using the Apache Spark job definition. You will be able to select the Python file that you have already uploaded to your Azure Data Lake Storage Gen2 account as part of the initial step. You can also upload the file at this stage and select it for the Spark job definition.

Command-line Arguments:

You have the option to pass arguments to your Apache Spark job if it is required. As it is optional, you can skip this as well. You can pass multiple command-line arguments to your job. You need to ensure that you separate each command-line argument by a space. In our case, the path of the data file and the path for the result are being passed as command-line arguments, which are separated by a space.

Reference Files:

You can specify the paths of any additional files you want to use in the main definition file. You also have the option to upload additional reference files to your storage account. In our scenario, we are using just one Python file, but if you have multiple Python files and your main Python file refers to those reference Python files, then you can upload those Python files using this option.

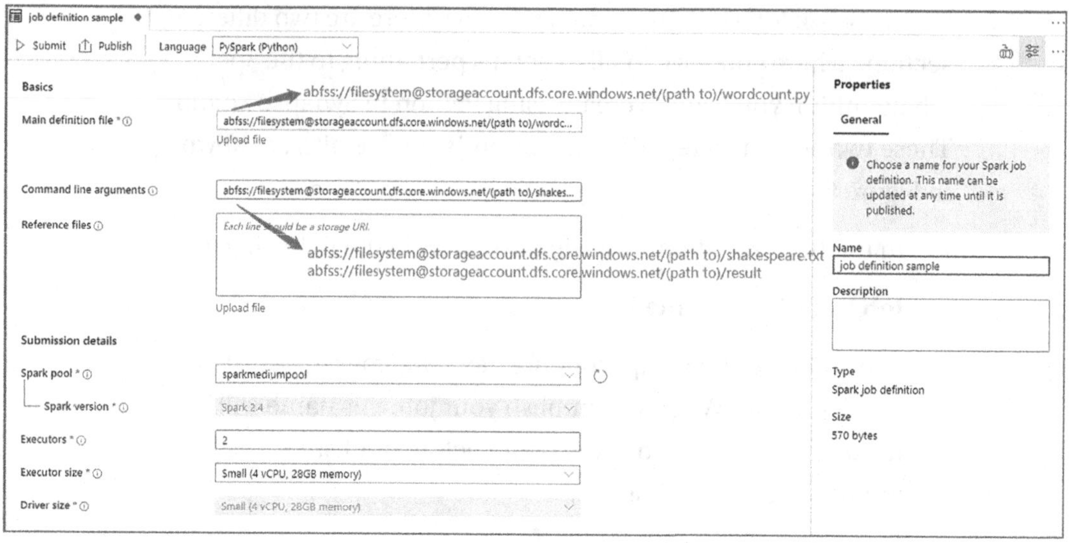

Figure 6-13. *Apache Spark job definition. Source: Microsoft*

- In the Submission Details section, you will have to fill in the following details:

Spark Pool:

Here, you get the option to select the Apache Spark Pool to which you want to submit your Apache Spark job definition. If you have multiple Apache Spark Pools in your Azure Synapse Analytics workspace, then you have to select one Apache Spark Pool from the dropdown, which will be used to execute your Apache Spark Job.

Spark Version:

This will show the Spark version being used by the Spark Pool, which you would have selected in the previous step from the dropdown. As it is an informational field, it is a read-only field.

Executors:

Enter the number of executors that you want to give to your Apache Spark job from the Apache Spark Pool on which you want to run your job. Your job will use whatever number of executors you have defined here. However, it cannot exceed the total executors available in the selected Apache Spark Pool.

Executor Size:

You can select the appropriate executor size from the dropdown. There are various executor sizes available in the dropdown, like small, medium, large, XLarge, and XXLarge. Executor sizes are available in combinations of cores and memory put together. For our example, you can select the small executor size, which will have four virtual CPUs and 28 GB of RAM or memory.

Driver Size:

This allows you to select the size of the Driver or Header node that will be used by your Apache Spark job definition. It will be dependent on your executor size, as generally driver size and executor size are kept the same.

- Once all the configurations are done for your Apache Spark job definition, you are ready to publish your job to the Apache Spark Pool. For that, you need to click the Publish button, as shown in Figure 6-14.

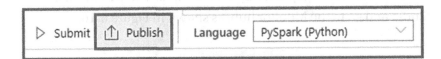

Figure 6-14. *Publish Apache Spark job definition. Source: Microsoft*

8. Once your Apache Spark job definition is published, you will be able to submit your job for execution to the Apache Spark Pool, as shown in Figure 6-15. Your Apache Spark job definition will be submitted to the Apache Spark Pool you configured in the definition file.

Figure 6-15. Submit Apache Spark job definition for execution. Source: Microsoft

How to Monitor Synapse Spark Pools Using Synapse Studio

If you have many Synapse Spark Pools, then it becomes difficult to monitor all of them and check which pool is using how many resources. Synapse Studio makes it easy to monitor all your Synapse Spark Pools through a GUI. You will be using Spark Pools to run notebooks, jobs, and other kinds of applications supported by Spark.

1. Log in to your Azure subscription via Azure Portal and go to your Azure Synapse Analytics workspace.

2. Within the Azure Synapse Analytics workspace, launch Synapse Studio and go to the Monitor hub, as shown in Figure 6-16.

Figure 6-16. Monitor hub in Synapse Studio. Source: Microsoft

3. Within the Monitor hub, under the Analytics Pools section, click on Apache Spark Pools, as shown in Figure 6-17.

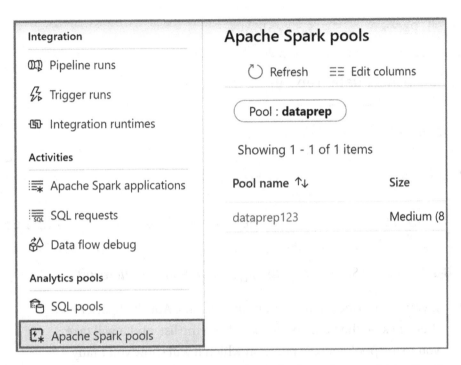

Figure 6-17. *Apache Spark Pools from Monitor hub under Analytics Pools. Source: Microsoft*

4. On the Apache Spark Pools page, you will be able to see the list of all the Apache Spark Pools that are available in your Azure Synapse Analytics workspace.

5. If you have many Apache Spark Pools in your Azure Synapse Analytics workspace and want to search quickly for a specific pool, then you also have an option to filter the list as per your requirements, as shown in Figure 6-18.

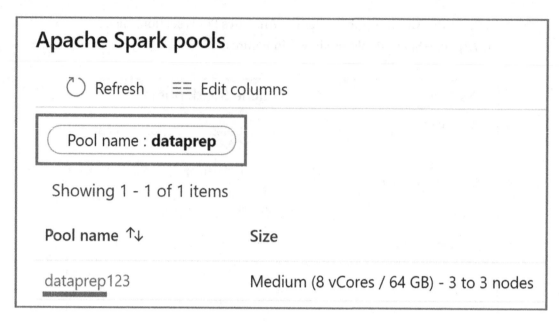

Figure 6-18. *Apache Spark Pools listing screen. Source: Microsoft*

6. If you want to look at details about a specific Apache Spark Pool, then click on the name of the pool from the list, and it will take you to the pool-specific page, on which it will show you many details about that specific Apache Spark Pool. This includes allocated virtual cores, allocated memory in GB, number of active applications on the pool, number of failed applications in last 15 mins, and so on, as shown in Figure 6-19. It will also show you which user is allocated how many virtual cores and a list of running applications, along with who submitted those applications.

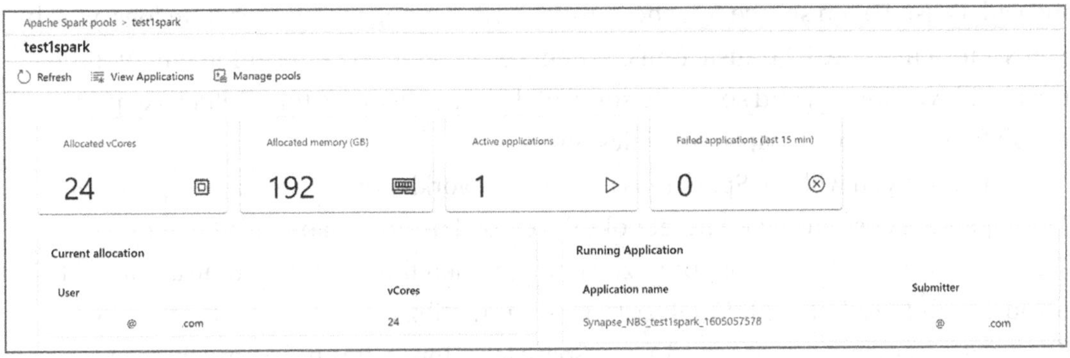

Figure 6-19. *Apache Spark Pool Monitoring screen. Source: Microsoft*

Summary

As mentioned at the start of the chapter, Apache Spark—or Synapse Spark, as we called it in this chapter—is one of the most important feature additions (on top of erstwhile Azure SQL Data Warehouse) in Azure Synapse Analytics. It brings much-needed real power and performance to Azure Synapse Analytics. Apache Spark is a de facto standard for any Big Data analytics workload. Synapse Spark is Microsoft's own implementation of Apache Spark within Azure Synapse Analytics. Synapse Spark can also be used for various machine learning workloads.

Apache Spark brings many features and capabilities to Azure Synapse Analytics. We discussed those features and capabilities in detail. Speed, faster start time, ease of creation, ease of use, automatic scalability, separation of concerns, multiple language support, integration with IDEs, pre-loaded libraries, and REST APIs are some of the important features and capabilities of Synapse Spark that we covered in some detail. These features and capabilities bring a lot of value to the table for any data analytics project. Synapse Spark is an important element to be considered for any workload you want to run using Azure Synapse Analytics.

Delta Lake is an open source storage layer that you can implement on top of your existing data lake implementation. It brings a lot of value when used in combination with Synapse Spark. It brings ACID transactions to your data lake, which are otherwise difficult to manage in data lakes. Apart from ACID transactions, Delta Lake brings many other benefits to Azure Synapse Analytics. Since it is tightly integrated with Synapse Spark, it is also every easy to use its features. Delta Lake tables can be used as source or sink for batch or streaming jobs out of the box. We also discussed that Delta Lake uses the widely used open source Apache Parquet format along with transaction logs, which helps it to achieve ACID transactions. Schema evaluation, schema enforcement, time travel with versioning, and so on are some of the other features that Delta Lake provides to you for your Azure Synapse Analytics projects.

Generally, you will use Spark jobs to run data workloads on a Synapse Spark Pool. These jobs are custom-coded notebooks developed using Synapse Studio in most scenarios. If your jobs are not optimized enough, then they may consume additional resources and result in additional costs. Optimum utilization of available resources like cores and memory is crucial for any Spark job. Hence, it is necessary that you understand certain aspects of Spark job optimizations. Therefore, we have discussed those aspects in detail, which include data format, memory management, data serialization, data caching, data abstraction, join and shuffle optimization, bucketing, and hyperspace indexing.

Azure Synapse Analytics provides many machine learning capabilities, and Synapse Spark plays an important role in providing those capabilities. Synapse Spark can be used appropriately for almost all stages of the data science process life cycle. Synapse Spark can help you in data preparation, data exploration, building machine learning models, training machine learning models, model deployment, and model scoring. This means that it can support an end-to-end data science process, which we discussed in detail.

Toward the end of the chapter, we discussed how to carry out certain important activities for Synapse Spark. We looked at a step-by-step process for how to create an Azure Synapse Spark Pool in an Azure Synapse Analytics workspace using Azure Portal. Later, we also looked at a detailed step-by-step process for submitting a Synapse Spark job definition to Synapse Spark Pool using Synapse Studio. We used a sample Python file for this exercise.

To summarize, we discussed Synapse Spark in some detail and covered many of its important aspects that may help you to understand Synapse Spark with its features, capabilities, optimization techniques, and so on. This will give you a good foundation on which to build your knowledge around Synapse Spark, as it is a very complex and vast topic. In the next chapter, we are going to discuss Synapse Pipelines in detail. Synapse Pipelines allows you to design, develop, and deploy your data pipelines visually through a graphical user interface. It is similar to Azure Data Factory in most respects. We will discuss this in more detail in the next chapter.

CHAPTER 7

Synapse Pipelines

For any data analytics project, you will likely need to ingest data from various disparate data sources and store it in a database, a data lake, a data warehouse, or a data lakehouse. To meet these requirements, you will have to build data ingestion pipelines, which will bring data to your desired target location. In addition, once you have ingested the data, you will have to cleanse it, apply business transformations and validations, and aggregate and consolidate it so that you can generate some insights and intelligence out of it. These processes require multiple jobs to be executed and orchestrated appropriately.

Traditionally, ETL/ELT tools are used to carry out data ingestion, data cleansing, transformation, aggregation, and so on. These tools allow you to connect various disparate data sources by using predefined connectors and apply transformations and aggregation functions, and later on you can write the transformed data back to your data sink, which also gets connected through a predefined connector. The Extract-Transform-Load, or ETL, process allows you to extract the data from data sources, transform the extracted data, and load the transformed data to a data sink. On the other side, the Extract-Load-Transform, or ELT, process allows you to extract the data from data sources and load it to the data sink, at which point the data is transformed to meet business requirements.

There is a plethora of proprietary as well as open source ETL/ELT tools available on the market. Due to the cloud's being well accepted, these tools have also started to appear in the cloud. Microsoft's SQL Server Integration Services (SSIS) is a classic example of this. SSIS is an on-premises service, and Microsoft also offers cloud-native Azure Data Factory (ADF), which is available as a Platform as a Service (PaaS) offering in the Azure cloud. ADF has native support for more than 90 connectors that can be used as data sources and data sinks for your data pipelines. Since it is a cloud-based service, it is also very easy and quick to get started with it. ADF provides a visual interface to design, develop, and deploy data pipelines. It is considered a low-code or no-code ETL tool. It also provides a comprehensive data pipeline monitoring facility.

© Bhadresh Shiyal 2021
B. Shiyal, *Beginning Azure Synapse Analytics*, https://doi.org/10.1007/978-1-4842-7061-5_7

Currently, you can use Azure Data Factory, Azure Data Lake Storage, Azure Databricks, and so forth as part of your solution, and you have to visit different screens within Azure to do so. However, with Azure Synapse Analytics, Microsoft has made all these services accessible under one tool, and that is Azure Synapse Studio. That means that for designing, developing, and deploying data pipelines, you need not go to Azure Data Factory separately, as you have that facility available within Synapse Studio. The Integrate hub within Synapse Studio will give you all the options you will need to develop synapse pipelines.

Synapse Pipelines is one of the core components of Azure Synapse Analytics. In Chapter 4, we discussed Synapse Pipelines at a very high level. In this chapter, we are going to take a deeper dive into it. We will start with an overview of Azure Data Factory, as that is the foundation for Synapse Pipelines, as almost all the main building blocks have been taken from it. Later, we will look at data movement activities, data transformation activities, and control flow activities. Toward the end of the chapter, we will discuss a couple of examples for copy pipelines as well as transformation pipelines and how to schedule a pipeline. So, let us get started.

Overview of Azure Data Factory

You will not be surprised that we are starting the chapter meant for Synapse Pipelines with an overview of Azure Data Factory, as we discussed in Chapter 4 that Synapse Pipelines is derived almost entirely from Azure Data Factory. So, it is better to understand Azure Data Factory before delving into Synapse Pipelines. Figure 7-1 contains various features of Azure Data Factory.

Microsoft introduced Azure Data Factory on October 28, 2014, in a public preview. It became generally available on August 6, 2015. It had only limited features at that point in time. SQL Server Integration Service was more feature rich as an integration tool than Azure Data Factory v1 at that time. However, things changed with the launch of Azure Data Factory v2. Looping, branching, visual drag and drop options, scheduling, a new GUI, and more were introduced in Azure Data Factory v2, and it was made generally available on June 27, 2018. By that time, Azure Data Factory had become a very mature product as a hybrid integration and orchestration tool. Microsoft introduced data flows into Azure Data Factory v2 in 2019, which added various capabilities to transform the data apart from just copying it. So, basically Azure Data Factory is a hybrid data integration and orchestration service that allows you to rapidly and proficiently design, develop, and deploy automated data pipelines without writing any code or in some circumstances very little code—a.k.a. low code. Thus, Azure Data Factory is a fully GUI-based no-code or low-code tool that will meet most of your data ingestion, transformation, and orchestration requirements out of the box.

INGEST	CONTROL FLOW	DATA FLOW	SCHEDULE	MONITOR
• **Multi-cloud and on-prem hybrid copy data** • **90+ native connectors** • **Serverless and auto-scale** • **Use wizard for quick copy jobs**	• Design code-free data pipelines • Generate pipelines via SDK • Utilize workflow constructs: loops, branches, conditional execution, variables, parameters, etc.	• Code-free data transformations that execute in spark • Scale-out with Azure Integration Runtimes • Generate Data Flows via SDK • Designers for data engineers and analysts	• Build and maintain operational schedule for your data pipelines • Wall-clock, event-based, tumbling windows, and chained	• View active executions and pipeline history • Detail activity and data flow executions • Establish alerts and notfications

Figure 7-1. *Azure Data Factory features*

Now, let us discuss the main features mentioned in the preceding figure. Data ingestion is one of the most important features as it allows you to connect disparate data sources and copy data from them. It supports over 90 native connectors that cover a wide array of most-used systems. You can copy data from not just the Azure cloud but also on-premises data stores as well as from other clouds. It provides a wizard-based GUI to connect and copy the data from various sources, which makes it easy and quick for anyone to set up those jobs as per their requirements.

As part of control flow, Azure Data Factory provides a feature to design code-free data pipelines. On top of that you also have the ability to generate a data pipeline via SDKs and APIs. Programmatically, you can create, modify, and deploy the data pipelines from your code as well. ADF also supports various data-driven workflow constructs, like loops, branches, conditional execution, variables, parameters, and so on. ADF also provides a facility to design data flows that can include your data transformation logic, which you can execute on a Spark cluster. The good part is that you can do it without writing any code. Because it's a Spark cluster, it will automatically scale out based on your workload, which eliminates your worries around spinning up and maintaining the cluster.

Once you have developed your control flow or data flow, you'll want to schedule its execution. ADF provides multiple options to do so. It allows you to schedule your workflow based on clock time. You can also take advantage of the event-based triggering mechanism that allows you to run your data pipeline based on certain predefined events, like the arrival of a new file in your blob storage. Once you have triggered your data pipeline, you will want to know what is happening at runtime in your workflow. That facility is also available, and you can see your active executions from the ADF interface itself very easily. It also allows you to look at historical data pipeline runs. You can also establish alerts and notifications for your pipeline runs, if required. Thus, Azure Data Factory is a comprehensive hybrid data integration and data orchestration tool that allows you to copy data from various data sources and provides the ability to transform the data using data flows, which can easily be scheduled or triggered based on events and can easily be monitored from the same interface.

Let us look at Synapse Pipelines now. We will start with an overview.

Overview of Synapse Pipelines

Azure Data Factory is the foundation for Synapse Pipelines. Instead of a separate instance of Azure Data Factory, you can have it as part of your Azure Synapse Analytics workspace. Synapse Pipelines is accessible from Synapse Studio, which you can launch from your workspace. In Synapse Studio, on the left-hand side, you have to click on the Integrate hub to launch the interface for Synapse Pipelines. As we already discussed the features of Azure Data Factory in the previous section, we are not going to discuss the features of Synapse Pipelines separately here, as they are the same.

Synapse Pipelines is a cloud-native data integration and data orchestration tool that provides you with ETL or ELT services based on how you want to utilize it. Like ADF, Synapse Pipelines allows you to design, develop, and deploy data-driven workflows to copy and transform data at scale. It can support really complex ETL or ELT processes, which can easily be developed using a GUI-based drag-and-drop feature without writing any code at all, or sometimes with a little code. The ability to design and develop data pipelines visually is the biggest advantage you have in Synapse Pipelines.

Now, we will look at some important concepts that are the building blocks for Synapse Pipelines and are derived from Azure Data Factory. We discussed this in brief in Chapter 4, but here we will go into more detail. You can refer to Figure 4-6 for this. We are going to discuss activities, pipelines, linked services, datasets, integration runtime (IR), parameters, and so forth in the following sections.

Activities

In Synapse Pipelines, activities define actions or tasks to be performed on your data. An activity will take zero or more inputs and produce one or more outputs. Generally, an activity can define a unique action or task that you want to perform on your data. If you want to perform multiple tasks or actions with your data pipelines, then you will have to use multiple different activities in your Synapse Pipelines. Activities are grouped together in different activity groups, as shown in Figure 7-2, which makes it easy to search for and locate any specific activity you want to use.

For example, a Databricks activities group consists of three different activities: Databricks notebook, Databricks Python, and Databricks JAR. A Databricks notebook activity will allow you to point to a Databricks notebook, which you can execute from Synapse Pipelines. Instead of a Databricks notebook, if you want to execute a Databricks JAR or a Databricks Python, then those activities are separately available as well.

There are also logical groupings of these activities that are not shown on the GUI. Logically, there are three different groups of activities, including data movement activities, data transformation activities, and control activities.

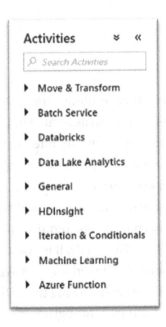

Figure 7-2. *Activities*

Pipelines

A pipeline is a logical grouping of activities put together to accomplish a task. In Synapse Pipelines, you may have one or more activities as part of your data pipeline. Synapse Pipelines also allow you to chain multiple activities inside a pipeline. For example, you can have a single data pipeline developed using Synapse Pipelines that allows you to ingest the data, cleanse the data, as well as transform the data. Here, you are going to chain or join three or more different activities together in a data pipeline.

Once you have designed and developed your data pipeline, you will be able to debug the pipeline to check if it is able to execute as per your requirements. You also have the option to set a breakpoint and debug the pipeline to a certain activity within your data pipeline. Once your pipeline meets your requirements, you can publish your changes to Synapse Pipelines. A published pipeline can be executed manually or on a trigger. Once you execute the pipeline, you will also be able to view pipeline runs, which will display what happened to your data pipeline while it was being executed. Historical pipeline runs can be viewed from the Synapse Pipelines GUI. Similarly, a run-time execution view of the pipeline is also visible, which shows you what is currently happening in your data pipeline.

Linked Services

Synapse Pipelines allows you to connect multiple disparate data sources as well as to define a different compute infrastructure so as to run your data pipeline through a core component called linked services. There are two different types of linked services that you can create in Synapse Pipelines: data store linked services and compute linked services.

The linked service that allows you to connect various data sources is known as a data store linked service. In simple words, a data store linked service is a connection string that contains details about various parameters required in order to connect to a specific data source using Synapse Pipelines. Generally, it is also referred to as a connector. There are more than 90 pre-defined data store linked services available in Synapse Pipelines. It is easy to define a data store linked service using the GUI. However, you also have the option to define data store linked services through code using any of the supported tools or SDKs, like .NET API, PowerShell, REST API, and Azure Resource Management (ARM) Template. To make it easy to understand, you can think of a data store linked service like a connection to a database server.

The second type of linked service is a compute linked service. As its name suggests, it allows you to define and configure the compute resources that you can use to execute your data pipeline. There are multiple compute infrastructure options available to you in a compute linked service. It includes on-demand or your own HDInsight Cluster, Azure Batch, Azure Machine Learning Studio (Classic), Azure Machine Learning, Azure Data Lake Analytics, Azure SQL Server, Azure Synapse Analytics, SQL Server, Azure Databricks, and Azure Functions. Based on your data pipeline activities, you will have to select the right type of compute linked services. It may happen that for a single data pipeline, you may end up using multiple different data store linked services as well as compute linked services.

Dataset

Data store linked services will allow you to connect various disparate data sources, but those data sources may contain multiple different tables, objects, or files. Data store linked services do not allow you to define which specific object, table, or file you need to choose for your data pipeline. That is where datasets come into the picture. They are another core component in Synapse Pipelines. A dataset represents the data structure within a specific data source that you can connect to using a data store linked service. A named view of data that simply points to or references the data you want to use in your data pipeline activities as inputs and outputs is known as a dataset in Synapse Pipelines. To make it easy to understand, you can compare a dataset to a table in a database.

We have discussed activities, pipelines, linked services, and datasets so far. It is important to understand the relationships between them. A pipeline is a logical grouping of activities that runs on compute linked services and consumes and produces datasets that represent a data item(s) stored in data store linked services. Does it sound confusing? If yes, study Figure 7-3 carefully, and you will be able to understand it easily.

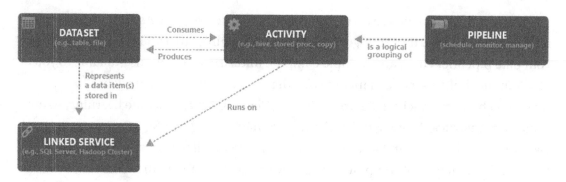

Figure 7-3. *Relationship among dataset, activity, pipeline, and linked service. Source: This image is taken from Microsoft Documentation titled "Linked services in Azure Data Factory" dated August 21, 2020.* `https://docs.microsoft.com/ en-us/azure/data-factory/concepts-linked-services`

Integration Runtimes (IR)

The integration runtime (IR) is the most important component of Synapse Pipelines. As its name suggests, it is the runtime that will be used by activities defined in your data pipelines. The integration runtime creates a bridge between the data sources with which you connect via linked services and the activities that you want to perform on the datasets derived from those data sources.

In simple terms, an integration runtime provides a compute infrastructure for the execution of your data pipelines. The integration runtime provides four main data integration capabilities: data flow, data movement, activity dispatch, and SSIS package execution. IR executes a data flow in a managed Azure compute infrastructure. It allows you to copy data across public and private networks, including on-premises. It also dispatches and monitors activities on various compute linked services. The integration runtime can also natively execute a legacy SSIS package in a managed Azure compute infrastructure. There are three different types of integration runtimes available to you. Let us discuss them one by one.

Azure Integration Runtime (Azure IR)

The most used integration runtime is Azure Integration Runtime. It provides data flow, data movement, and activity dispatch capabilities in public as well as private networks. It is widely used because it provides three out of four main IR capabilities. It is a fully managed and serverless compute environment. That means that you do not have to take care of infrastructure, software installation, patching, or scaling. All these aspects are taken care of by Microsoft. It is a pay-as-you-go model, so you pay only for the duration of actual utilization. Azure IR also supports scaling compute using data integration units (DIU). This is a measure to represent the power of a single unit, which combines CPU, memory, and network resources. Azure IR scales up automatically based on set DIUs.

Azure IR will allow you to run data flows in Azure and run copy activity between cloud data stores. It also supports dispatch activity functionality for many of the activities in the public network, which include activities related to Databricks, HDInsight, Azure Machine Learning, and more.

Azure IR allows you to connect disparate data sources and different compute services through publicly accessible service endpoints. However, you also have the option to connect to those data stores using a private link service on a private network. This option is in preview at the time of writing this chapter. You can find more details about it at https://docs.microsoft.com/en-us/azure/data-factory/managed-virtual-network-private-endpoint.

Self-Hosted Integration Runtimes (SHIR)

The second type of integration runtime is Self-Hosted Integration Runtime (SHIR). As its name suggests, it is self-hosted, which means that you will have to provide a hosting infrastructure for this integration runtime. You will also have to install and manage the integration runtime on the hosting infrastructure. Java Runtime Environment (JRE) is a core dependency for Self-Hosted Integration Runtime, so it has to be installed on the same host on which you are going to install the Self-Hosted Integration Runtime executables.

The SHIR provides data movement and activity dispatch capabilities in private as well as public networks. However, it does not support data flow capability as of now. Generally, Self-Hosted Integration Runtime is preferred when you want to integrate various data sources securely in a private network infrastructure. You can install Self-Hosted Integration Runtime executables on an on-premises environment behind your corporate network firewall. You can also install it inside a private virtual network.

The important point to note here is that Self-Hosted Integration Runtime only makes outbound HTTP-based connections to the open internet, which means that you cannot use those linked services that rely on non-HTTP-based protocols like FTP for connecting via the internet from the Self-Hosted Integration Runtime. Additionally, if you have a proxy server set up in your environment and you want your Self-Hosted Integration Runtime to connect via HTTP proxy, that option is available to you in SHIR. You can have multiple nodes in your Self-Hosted Integration Runtime, but currently a maximum of four nodes are supported. Another limitation to remember is that all these nodes should be running Windows OS only, as SHIR does not support other OSes like Linux, etc.

Azure SSIS Integration Runtimes (Azure SSIS IR)

The third type of integration runtime is Azure SSIS Integration Runtime. This provides the capability to natively execute legacy SSIS packages from Azure Data Factory only, as it is not supported in Synapse Pipelines. You can connect on-premises data sources by connecting your Azure SSIS IR to a virtual network that is connected with your on-premises network. Azure SSIS Integration Runtime is a fully managed cluster of Azure VMs that are solely dedicated to running your legacy SSIS packages.

Since your legacy SSIS packages are going to get executed from your Azure Data Factory interface, it needs to have access to those packages from cloud storage. Hence, you will have to publish your SSIS packages to either Azure SQL Database or SQL Managed Instance. Alternatively, you can also store these packages to a file system like Azure Files. You also have the option to bring your Azure SSIS IR–hosted Azure VMs into your Azure Virtual Network, which can give you better security controls on your Azure SSIS IR. You will have to carefully evaluate data integration capabilities and network support for each of these integration runtimes to decide which one will meet your requirements.

Control Flow

You can include multiple different activities in a single data pipeline in Synapse Pipelines. That means that you will need an orchestration mechanism that allows you to control the execution flow of the various activities included in your data pipeline. The orchestration of various pipeline activities is known as control flow in Synapse Pipelines. Control flow allows you to achieve sequencing, chaining, looping, branching, and parameterizing of your pipeline activities. We will look at the list of control flow activities that you can use in your control flow data pipeline later in this chapter.

Parameters

You might like to change the flow of your pipeline based on variation in your inputs. You can achieve this by using parameters in Synapse Pipelines. By using parameters in your data pipeline, you enable your data pipeline to make dynamic decisions based on the values you pass to your parameters. Parameter values are consumed by different activities in your data pipeline. A parameter is a key–value pair that is read-only and cannot be changed during the execution of the pipeline.

Data Flow

Data flows are supported in Synapse Pipelines as well, which is known as mapping data flow in Azure Data Factory. Basically, data flows are visually designed data transformations in Synapse Pipelines. Data flows allow you to design and develop data transformation logic visually without writing any code. Synapse Pipelines uses scaled-out Apache Spark clusters to execute the data flows. Data flow activities can utilize the existing capabilities from Synapse Pipelines, like visual authoring, debugging, scheduling, monitoring, and control flow.

We have discussed various important concepts for Synapse Pipelines in detail in the preceding sections. Activities within Synapse Pipelines have a lot of varieties and provide multiple capabilities to your Synapse Pipelines. Let us discuss some of the important activities in the following sections. These activities are grouped into three different groups, and we are going to discuss each one separately. So, let us start with data movement activities, since that is the core for any data integration workload.

Data Movement Activities

In the data movement activities group, you have a copy activity, which supports multiple diverse source data stores and sink data stores. The copy activity is supported by Azure Integration Runtime as well as Self-Hosted Integration Runtime. In Synapse Pipelines, a copy activity will allow you to copy data from a source data store to a sink data store. It plays an important role when you want to ingest data from various disparate data sources to your target data store.

There are multiple data stores that are supported by copy activity in Synapse Pipelines. Due to the long list of supported data stores, these are divided into different categories and are tabulated in Table 7-1, Category: Azure; Table 7-2, Category: Database; Table 7-3, Category: NoSQL; Table 7-4, Category: File; Table 7-5, Category: Generic, and Table 7-6, Services and Applications. Microsoft Documentation is the source for all of these tables.

Category: Azure

Table 7-1. Category: Azure

Data Store	Supported as a Source	Supported as a Sink	Supported by Azure IR	Supported by Self-Hosted IR
Azure Blob storage	✓	✓	✓	✓
Azure Cognitive Search index		✓	✓	✓
Azure Cosmos DB (SQL API)	✓	✓	✓	✓
Azure Cosmos DB's API for MongoDB	✓	✓	✓	✓
Azure Data Explorer	✓	✓	✓	✓
Azure Data Lake Storage Gen1	✓	✓	✓	✓
Azure Data Lake Storage Gen2	✓	✓	✓	✓
Azure Database for MariaDB	✓		✓	✓
Azure Database for MySQL	✓	✓	✓	✓
Azure Database for PostgreSQL	✓	✓	✓	✓
Azure Databricks Delta Lake	✓	✓	✓	✓
Azure File Storage	✓	✓	✓	✓
Azure SQL Database	✓	✓	✓	✓
Azure SQL Managed Instance	✓	✓	✓	✓
Azure Synapse Analytics	✓	✓	✓	✓
Azure Table Storage	✓	✓	✓	✓

Category: Database

Table 7-2. *Category: Database*

Data Store	Supported as a Source	Supported as a Sink	Supported by Azure IR	Supported by Self-Hosted IR
Amazon Redshift	✓		✓	✓
DB2	✓		✓	✓
Drill	✓		✓	✓
Google BigQuery	✓		✓	✓
Greenplum	✓		✓	✓
HBase	✓		✓	✓
Hive	✓		✓	✓
Apache Impala	✓		✓	✓
Informix	✓	✓		✓
MariaDB	✓		✓	✓
Microsoft Access	✓	✓		✓
MySQL	✓		✓	✓
Netezza	✓		✓	✓
Oracle	✓	✓	✓	✓
Phoenix	✓		✓	✓
PostgreSQL	✓		✓	✓
Presto	✓		✓	✓
SAP Business Warehouse via Open Hub	✓			✓
SAP Business Warehouse via MDX	✓			✓
SAP HANA	✓	✓		✓
SAP table	✓			✓

(*continued*)

Table 7-2. (*continued*)

Data Store	Supported as a Source	Supported as a Sink	Supported by Azure IR	Supported by Self-Hosted IR
Snowflake	✓	✓	✓	✓
Spark	✓		✓	✓
SQL Server	✓	✓	✓	✓
Sybase	✓			✓
Teradata	✓		✓	✓
Vertica	✓		✓	✓

Category: NoSQL

Table 7-3. *Category: NoSQL*

Data Store	Supported as a Source	Supported as a Sink	Supported by Azure IR	Supported by Self-Hosted IR
Cassandra	✓		✓	✓
Couchbase (Preview)	✓		✓	✓
MongoDB	✓		✓	✓
MongoDB Atlas	✓		✓	✓

Category: File

Table 7-4. *Category: File*

Data Store	Supported as a Source	Supported as a Sink	Supported by Azure IR	Supported by Self-Hosted IR
Amazon S3	✓		✓	✓
File system	✓	✓	✓	✓

(*continued*)

Table 7-4. (*continued*)

Data Store	Supported as a Source	Supported as a Sink	Supported by Azure IR	Supported by Self-Hosted IR
FTP	✓		✓	✓
Google Cloud Storage	✓		✓	✓
HDFS	✓		✓	✓
SFTP	✓	✓	✓	✓

Category: Generic

Table 7-5. *Category: Generic*

Data Store	Supported as a Source	Supported as a Sink	Supported by Azure IR	Supported by Self-Hosted IR
Generic HTTP	✓		✓	✓
Generic OData	✓		✓	✓
Generic ODBC	✓	✓		✓
Generic REST	✓	✓	✓	✓

Category: Services and Applications

Table 7-6. *Category: Services and applications*

Data Store	Supported as a Source	Supported as a Sink	Supported by Azure IR	Supported by Self-Hosted IR
Amazon Marketplace Web Service	✓		✓	✓
Common Data Service	✓	✓	✓	✓
Concur (Preview)	✓		✓	✓
Dynamics 365	✓	✓	✓	✓

(*continued*)

Table 7-6. (*continued*)

Data Store	Supported as a Source	Supported as a Sink	Supported by Azure IR	Supported by Self-Hosted IR
Dynamics AX	✓		✓	✓
Dynamics CRM	✓	✓	✓	✓
Google AdWords	✓		✓	✓
HubSpot	✓		✓	✓
Jira	✓		✓	✓
Magento (Preview)	✓		✓	✓
Marketo (Preview)	✓		✓	✓
Microsoft 365	✓		✓	✓
Oracle Eloqua (Preview)	✓		✓	✓
Oracle Responsys (Preview)	✓		✓	✓
Oracle Service Cloud (Preview)	✓		✓	✓
PayPal (Preview)	✓		✓	✓
QuickBooks (Preview)	✓		✓	✓
Salesforce	✓	✓	✓	✓
Salesforce Service Cloud	✓	✓	✓	✓
Salesforce Marketing Cloud	✓		✓	✓
SAP Cloud for Customer (C4C)	✓	✓	✓	✓
SAP ECC	✓		✓	✓
ServiceNow	✓		✓	✓
SharePoint Online List	✓		✓	✓
Shopify (Preview)	✓		✓	✓
Square (Preview)	✓		✓	✓
Web Table (HTML Table)	✓			✓
Xero	✓		✓	✓
Zoho (Preview)	✓		✓	✓

You can see from the preceding tables that Synapse Pipelines supports many of the Azure and non-Azure data stores natively. While looking for support for any of these data stores, you need to ensure that you check based on whether the data store is supported for your requirements or not. For example, there could be a scenario where a data store is supported only as a data source or data sink, while your requirements may be opposite to that. Similarly, you will have to pay attention to the integration runtime it supports and whether you have the same integration runtime available for your project.

Data Transformation Activities

Data transformation activities are those activities that can apply business transformations to your data within your Synapse Pipelines. A data transformation activity gets executed in a compute environment. You can either use a data transformation activity individually or chain it with another activity. Azure Databricks, Azure HDInsight (Hadoop), Azure SQL Server, Azure Synapse Analytics, SQL Server, Azure VM, Azure Batch, Azure Functions, and Azure Data Lake Analytics are the supported compute environment for it. Table 7-7 contains the list of data transformation activities.

Table 7-7. *Data Transformation Activities*

Data Transformation Activity	Compute Environment
Data Flow	Apache Spark Clusters managed by Azure Synapse Analytics
Azure Function	Azure Functions
Hive	HDInsight [Hadoop]
Pig	HDInsight [Hadoop]
MapReduce	HDInsight [Hadoop]
Hadoop Streaming	HDInsight [Hadoop]
Spark	HDInsight [Hadoop]
Azure Machine Learning Studio (classic) activities: Batch Execution and Update Resource	Azure VM

(continued)

Table 7-6. (*continued*)

Data Transformation Activity	Compute Environment
Stored Procedure	Azure SQL, Azure Synapse Analytics, or SQL Server
U-SQL	Azure Data Lake Analytics
Custom Activity	Azure Batch
Databricks Notebook	Azure Databricks
Databricks Jar Activity	Azure Databricks
Databricks Python Activity	Azure Databricks

Control Flow Activities

A control flow activity will allow you to control the flow of your data pipeline. If you are coming from a programming background, then you will realize that most of the control flow activities are similar to some of the widely used programming language constructs. Based on the requirements of your business logic, you need to ensure that you pick the right type of activity for your data pipeline. Table 7-8 contains the list of control flow activities.

Table 7-8. *Control Flow Activities*

Control Activity	Description
Append Variable	Add a value to an existing array variable.
Execute Pipeline	Execute Pipeline activity allows a Synapse Pipelines to invoke another pipeline.
Filter	Apply a filter expression to an input array.
For Each	For Each activity defines a repeating control flow in your pipeline. This activity is used to iterate over a collection and executes specified activities in a loop. The loop implementation of this activity is similar to the Foreach looping structure in programming languages.
Get Metadata	Get Metadata activity can be used to retrieve metadata of any data in Synapse Pipelines.

(*continued*)

Table 7-8. (*continued*)

Control Activity	Description
If Condition Activity	The If Condition can be used to branch based on a condition that evaluates to true or false. The If Condition activity provides the same functionality that an `if` statement provides in programming languages. It evaluates a set of activities when the condition evaluates to true and another set of activities when the condition evaluates to false.
Lookup Activity	Lookup activity can be used to read or look up a record/table name/value from any external source. This output can further be referenced by succeeding activities.
Set Variable	Set the value of an existing variable.
Until Activity	Implements Do-Until loop that is similar to `Do-Until` looping structure in programming languages. It executes a set of activities in a loop until the condition associated with the activity evaluates to true. You can specify a timeout value for the Until activity in Synapse Pipelines.
Validation Activity	Ensure a pipeline only continues execution if a reference dataset exists, meets a specified criteria, or a timeout has been reached.
Wait Activity	When you use a Wait activity in a pipeline, the pipeline waits for the specified time before continuing with execution of subsequent activities.
Web Activity	Web activity can be used to call a custom REST endpoint from a Synapse Pipelines. You can pass datasets and linked services to be consumed and accessed by the activity.
Webhook Activity	Using the Webhook activity, call an endpoint and pass a callback URL. The pipeline run waits for the callback to be invoked before proceeding to the next activity.

Copy Pipeline Example

A copy pipeline will include a copy activity in it, which will allow you to connect source and sink data stores using Synapse Pipelines so that you can copy data from the source to the sink automatically. The copy activity is part of the data movement activities and supports multiple diverse data sources natively. We studied the list of these data stores while discussing the data transformation activities category. Internally, a copy activity inside a copy pipeline performs the following steps in the given sequence:

1. Connects and reads data from the source data store.

2. Performs serialization/deserialization, compression/ decompression, column mapping, and so on. The copy activity will perform these actions based on the configuration of your input dataset as well as output dataset and copy activity configuration.

3. Write data to sink or destination data store as per configuration.

If you want to copy data from source to destination without making any changes to its format, then the copy activity will be able to do that very easily and efficiently as it will not have to carry out serialization and deserialization tasks during the copying exercise. However, the copy activity supports copying files and converting them from one format to another format. The copy activity supports the following eight file formats in Synapse Pipelines:

1. Excel

2. XML

3. JSON

4. Delimited Text

5. Parquet

6. Avro

7. Binary

8. ORC

Now, let us look at the stepwise process of creating a copy activity data pipeline in Synapse Pipelines.

1. Create the linked service for the source data store. You can look at the list of available data stores that can be used as the source in the "Data Movement Activities" section to check if your desired data store is supported as a source data store.

2. Similarly, create the linked service for the destination or sink data store. For that as well, you can check the "Data Movement Activities" section to see if your required data store is supported as a sink.

3. Create a data pipeline with the copy activity in Synapse Pipelines.
    For that you can drag and drop the copy activity on the canvas.
    Then complete the necessary configuration steps to set up the
    source and sink by selecting the previously created linked services
    for source and sink respectively from the GUI.

4. You can publish your pipeline and run it in debug mode to see
    what is happening when you execute your copy activity pipeline.
    Synapse Pipelines provides a GUI as well as programmatic options
    to easily monitor your data pipeline.

Transformation Pipeline Example

There are around 15 different data transformation activities that you can use in your
Synapse Pipelines. Based on the data transformation activities you choose for your data
pipeline, it will get executed on a different compute environment as tabulated in the
"Data Transformation Activities" section.

We will look at how to use a stored procedure as a data transformation activity
for this transformation pipeline example. As mentioned in the table in the "Data
Transformation Activities" section, a stored procedure activity will require one of the
following as the compute environment:

1. Azure SQL Database

2. Azure Synapse Analytics

3. SQL Server Database

For SQL Server Database, you will have to use Self-Hosted Integration Runtime on
the same machine on which your SQL Server Database is hosted. You can also connect
to a separate machine, provided the SHIR has access on that SQL Server Database.

Now, let us look at the high-level steps to design a data transformation pipeline:

1. Create a compute linked service, which will provide the compute
    environment for your data transformation pipeline. Use Azure
    SQL Database to register a compute linked service for this
    purpose.

2. Drag and drop a stored procedure activity on the canvas in the Synapse Pipelines GUI interface using Synapse Studio.

3. Configure Activity Type as "SqlServerStoredProcedure" in the configuration pane.

4. In the linked service, select the compute linked service you created in Step 1.

5. In the Stored Procedure Name field, specify the stored procedure you want to invoke for your data transformation pipeline. This stored procedure will contain the business transformation logic you want to apply on the data to transform it.

6. Optionally, you can also specify stored procedure parameters if there are any.

7. Publish your data transformation pipeline and debug to check if it is working as per your requirements.

8. Once the data transformation pipeline completes the execution, you can verify if the stored procedure activity has applied the required business transformations on the data by querying the data.

Pipeline Triggers

You can execute the data pipelines in Synapse Pipelines either manually or by using a trigger. A manual pipeline trigger is also known as on-demand execution as it allows you to execute the pipeline whenever you want to execute it as per your demand.

With a manual trigger, you have multiple options to trigger the data pipeline. You can use any one of the following options to manually trigger the data pipeline:

1. GUI from Synapse Studio

2. Python SDK

3. .NET SDK

4. REST API

5. Azure PowerShell module

Out of these, the easiest way to trigger a pipeline manually is obviously through the GUI from Synapse Studio, as you will not have to write any code, while with any other option, you will have to write some code. Apart from manually triggering the data pipeline, you also have the option to run the data pipeline through triggers. There are three different types of triggers supported by Synapse Pipelines, as follows:

1. Schedule trigger

2. Tumbling window trigger

3. Event-based trigger

A schedule trigger will allow you to run your pipeline based on a wall-clock schedule. A tumbling window trigger is a type of trigger that can fire at a periodic time interval from a specified start time, and it also allows you to retain the state of your data pipeline. An event-based trigger is a type of trigger that will get fired as a response to certain events. For example, if you want to run a data pipeline when a new file arrives or an existing file gets deleted from Azure Blob Storage, then you can use an event-based trigger for your data pipeline. It allows you to specify the path to watch for any new arrival or deletion based on your configuration.

Summary

To make Azure Synapse Analytics a comprehensive analytics platform, Microsoft has included Synapse Pipelines as a built-in integration and orchestration tool. Synapse Pipelines is built entirely on the strong foundation of Azure Data Factory (ADF), which is a proven and widely used cloud-native data integration and orchestration service. Integrating myriad and disparate data sources and ingesting the data into a data lake or data warehouse is a real challenge that any data analytics platform will have to deal with appropriately. By embedding Synapse Pipelines into Azure Synapse Analytics, Microsoft has achieved that goal.

Azure Data Factory has multiple building blocks that are almost the same as those in Synapse Pipelines. So, after presenting an overview of Azure Data Factory, we went on to discuss Synapse Pipelines and looked at each of those building blocks. Activities, pipelines, datasets, integration runtime, linked services, data flow, control flow, parameters, and so on are the basic building blocks of Synapse Pipelines.

Activities in Synapse Pipelines are the core building block in which a major chunk of work happens. We looked at three different groups of activities in tabular format. We examined data movement activities, data transformation activities, and data flow activities. Data movement activities include the copy activity, and Synapse Pipelines supports around 90 different data stores as the source and sink. These are connectors that are readily available in Synapse Pipelines and make it easy to connect to widely used data stores using the GUI.

Data transformation activities that allow you to apply various business transformations on your data using Synapse Pipelines can get executed on different compute environments. Data transformation activities support Azure HDInsight, Azure Functions, Azure VM, Azure Batch, Azure SQL Database, Azure Synapse Analytics, Apache Spark, SQL Database, and Azure Data Lake Analytics as the compute environment. We also discussed the control flow activity, which allows you to control the flow of your data pipeline. A few of the control flow activities are similar to some programing constructs, like `if`, `for` loop, `until` activity, and others.

Toward the end of the chapter, we discussed how to create an example data pipeline using a copy activity. We also looked at how to create an example data pipeline using a data transformation activity named store procedure, which requires a compute environment to execute your data pipeline. We also looked at how you can run your data pipeline. There are two options: manual or trigger-based. The trigger-based execution of a pipeline opens up many possibilities for scheduling or firing your data pipeline upon a specific event.

This chapter has provided a detailed understanding of Synapse Pipelines and its core components. This will help you to understand which connector is available in Synapse Pipelines and what the important activities are that you can include in your data pipeline. In the next chapter, we are going to discuss Synapse Studio and its core components in detail. So, stay tuned!

CHAPTER 8

Synapse Workspace and Studio

In previous chapters, we saw that Azure Synapse Analytics consists of multiple tools and technologies. Its architecture is also a little complex due to the amalgamation of many tools and technologies in it. For example, it contains three different compute engines. It includes Azure Data Factory, which is an entirely independent Azure Service, as Synapse Pipelines. It also allows you to integrate Power BI reports. It allows you to connect to multiple data stores. So, apart from the components it inherited from Azure SQL Data Warehouse, it has a lot of new components in it to make it a comprehensive and integrated data analytics platform.

To aid in the provisioning and management of all these components, Microsoft introduced Synapse workspaces to Azure Synapse Analytics. In Azure, a workspace is not a new concept, as there are many Azure services with workspaces. Basically, a workspace is a dedicated area in which to work with that service in Azure. A workspace also provides a security boundary to secure your service area. There are multiple ways in which you can create a Synapse workspace. The workspace is the central place from which you will interact with Azure Synapse Analytics. Along with your Azure Synapse Analytics workspace, you also get Synapse Studio. It is a central place through which you can carry out multiple activities in Azure Synapse Analytics.

In this chapter, we are going to cover the Azure Synapse Analytics workspace. We will discuss its main components and features. Azure Synapse Studio is a core component in the Synapse workspace, so we are going to discuss it in more detail. We will cover its main features and capabilities in detail. Azure Synapse Studio also allows you to connect with Power BI workspace. This is an interesting capability that we will cover in this chapter as it allows you to design your Power BI reports without leaving the Synapse workspace. So, let us get started with the Synapse workspace first, followed by Synapse Studio.

© Bhadresh Shiyal 2021
B. Shiyal, *Beginning Azure Synapse Analytics*, https://doi.org/10.1007/978-1-4842-7061-5_8

What Is a Synapse Analytics Workspace?

If you have used other Azure services, then you should be familiar with the workspace concept. Generally, a workspace is a central location on Azure Portal where you can carry out many of the tasks or actions pertaining to a specific Azure service. A workspace will make it possible to easily access all the resources and related information for a specific Azure Service from a dashboard-like home screen for that specific service.

Similarly, the central location from which you can access various resources pertaining to Azure Synapse Analytics, like storage, Serverless SQL Pools, and so forth; can view information about these resources; and can connect to these resources from the Azure Portal is known as the Azure Synapse Analytics workspace. The initial setup looks similar to Figure 8-1.

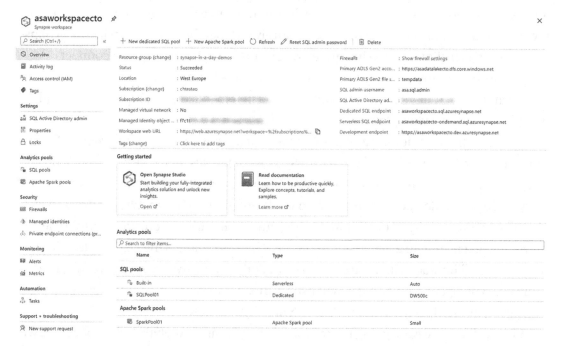

Figure 8-1. *Azure Synapse Analytics workspace. Source: Microsoft*

In Microsoft's words, a Synapse workspace is a securable collaboration boundary for doing cloud-based enterprise analytics in Azure. A workspace is deployed in a specific region and has an associated ADLS Gen2 account and file system for storing temporary data. A workspace is under a resource group. A workspace allows you to perform analytics with SQL and Apache Spark. Resources available for SQL and Spark analytics are organized into SQL and Spark pools.

If you want to provision or deploy Azure Synapse Analytics in your Azure subscription, then the very first step would be to create an Azure Synapse Analytics workspace. You can create one using different methods, which include Azure Portal, Azure CLI, Azure PowerShell, and Azure Resource Manager (ARM) template. Obviously, creating an Azure Synapse Analytics workspace using Azure Portal will be the easiest method compared to the other methods as it is completely based in the graphical user interface (GUI).

Synapse Analytics Workspace Components and Features

Azure Synapse Analytics consists of multiple components, and the Synapse workspace is one of them. The Azure Synapse Analytics workspace consists of multiple components as well. When you provision or create an Azure Synapse Analytics workspace in your Azure subscription, it will also create certain additional components, which include an Azure Data Lake Storage Gen2 account and a file system. It will also create a Serverless or On-Demand Synapse SQL Pool for you. On the Azure Synapse Analytics workspace's overview page, you will be able to see Serverless Synapse SQL Pool's end points.

Azure Data Lake Storage Gen2 Account and File System

The Azure Synapse Analytics workspace provisioning process requires you to specify an Azure Data Lake Storage Gen2 account. Along with this account, it will also ask you to provide a file system name from your Azure Data Lake Storage Gen2 account which you have already specified. Alternatively, you can create a new file system if one does not exist on the specified Azure Data Lake Storage Gen2. These are mandatory requirements, as these details will be used by Azure Synapse Analytics to decide the default location for storing logs and job output.

Serverless Synapse SQL Pool

Along with provisioning an Azure Synapse Analytics workspace, a Serverless or On-Demand Synapse SQL Pool will also get provisioned for you. On the overview page of your Azure Synapse Analytics workspace, you will be able to see the endpoint for the Serverless Synapse SQL Pool. Using that endpoint, you will be able to query the data immediately. You need not to provision this SQL pool separately as it is automatically provisioned for you. However, if you want to use a Dedicated or Provisioned Synapse SQL Pool, then you will have to provision it separately. We discussed the Serverless Synapse SQL Pool in detail in Chapter 5.

Shared Metadata Management

The Azure Synapse Analytics workspace enables different computational engines like Synapse Spark Pool and Serverless Synapse SQL Pool to share databases and tables. For example, if you have created any databases and Parquet tables using Synapse Spark Pools, then those databases and Parquet tables will also be available in a Serverless Synapse SQL Pool automatically. Using a Synapse Spark Pool, you can create databases, external tables, managed tables, and views. You will be able to access and query the databases, external tables, and managed tables because metadata is shared by the Azure Synapse Analytics workspace using the Serverless Synapse SQL Pool. However, views created by Synapse Spark Pool will not be available for querying by the Serverless Synapse SQL Pool, as views require a Spark engine to process the defining Spark SQL statement. Therefore, Spark views are shared among Spark pools only and not with SQL pools.

The Shared Metadata Management feature only shares the metadata of databases and tables across Synapse Spark Pools and Serverless Synapse SQL Pools, with some limitations. The first limitation is that you cannot make any updates or deletes using a Serverless SQL Pool in the databases and tables created using a Synapse Spark Pool. That means that you can query them but cannot make any changes in them. Another limitation is that the changes made in databases and tables by a Synapse Spark Pool will be reflected in the Serverless Synapse SQL Pool after some time. There is a short delay as the synchronization between Synapse Spark Pools and Serverless Synapse SQL Pools is done asynchronously.

Now, let us check how the security model works in the case of shared metadata. A basic thing to remember here is that the databases and tables are stored in Azure Data Lake Storage Gen2, and, because of that, the security model the Azure Synapse Analytics workspace follows is also at the storage level. That means that the permissions that are given at the folder and file levels in storage will be applicable for both Serverless Synapse SQL Pools and Synapse Spark Pools. For example, if you create an external table with authentication pass-through enabled on it, the data is secured for this table at the folder and file levels. If a user queries this external table, then the security identity of that user will be passed through to the file system, and if that user has access to that folder and files then that user will be able to access the external table; otherwise, it will not be accessible.

Code Artifacts

Code artifacts are the collections of code which you write in Azure Synapse Analytics workspace. The Azure Synapse Analytics workspace provides different options to write your code based on what you are trying to achieve and which computing engine you are targeting for your code. Using Synapse Studio, you can write a Spark notebook that can be executed against a Synapse Spark Pool. You can create, develop, and run notebooks using Synapse Studio within the Azure Synapse Analytics workspace. You can also select the primary coding language out of four available options, which include pySpark (Python), Spark(Scala), Spark SQL, and Spark .NET (C#). You can also use multiples of these languages in your Synapse Spark notebook by using the appropriate magic commands. The notebook code will allow you to load data from Azure Blob Storage, Azure Data Lake Storage Gen2, and Synapse SQL Pools.

The Azure Synapse Analytics workspace also provides the facility to write SQL scripts through Synapse Studio. It also allows you to import an existing SQL script into your workspace. The workspace also gives you the option to select which SQL pool you want to use to run your SQL script from a dropdown. Similarly, you have an option to select which database you want to use for your SQL script. There is also a facility to export the SQL script results to CSV, Excel, JSON, or XML format based on your requirements. You can also visualize your SQL script results using charts. You can select any chart type from the available options that suits your SQL script output. Generated chart output can also be saved as an image. You can also use a SQL script to explore data from a Parquet file stored in a storage account in an Azure Synapse Analytics workspace.

Apart from Spark notebooks and SQL scripts, the Azure Synapse Analytics workspace also provides the facility to design and develop data flows using Synapse Studio. Data flows are visually designed data transformations developed without writing any code. Data flows are executed on Apache Spark–based scaled out clusters as part of Synapse Pipelines. Through Synapse Studio, the Azure Synapse Analytics workspace provides facilities to author, debug, deploy, and monitor data flows natively.

Thus, as part of the code artifact feature, the Azure Synapse Analytics workspace provides various options and features for authoring, running, and monitoring Spark notebooks, SQL scripts and data flows. As we have discussed, all these options and features are available to use from Synapse Studio, which is part of the Azure Synapse Analytics workspace. Now, let us discuss Synapse Studio in more detail as that is one of the most important components of the Azure Synapse Analytics workspace.

What Is Synapse Studio?

Synapse Studio is the heart of the Azure Synapse Analytics workspace. It is a sort of one-stop shop for all your needs. It is the primary and core tool that allows you to interact with many of the components available to you in Azure Synapse Analytics. In simple words, it is a cloud-native web-based GUI tool that provides you with a central place from which you can carry out all the types of tasks and activities for the Azure Synapse Analytics workspace.

Azure Synapse Analytics includes multiple tools and technologies so as to offer a comprehensive data analytics experience. Thus, it becomes necessary to have a central place where all these tools and technologies can be accessed easily and efficiently so the end users can rapidly create, design, develop, debug, run, and deploy various solutions. There are mainly four different workloads in Azure Synapse Analytics, which include control flow, data flow, SQL scripts, and Spark notebooks. As we discussed in the previous section, Azure Synapse Studio provides easy and efficient options to develop, debug, and deploy all these workloads natively.

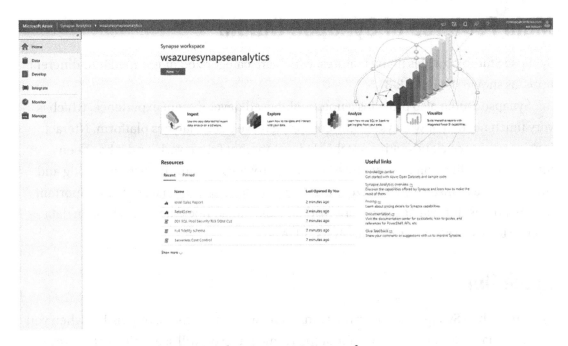

Figure 8-2. *Azure Synapse Studio. Source: Microsoft*

Azure Synapse Studio follows a hub-based navigation pattern to group various core functionalities it provides to the end user. These groups are called hubs, as seen on the left-hand side of the Azure Synapse Studio UI. Figure 8-2 shows the home hub as the selected hub. We will discuss each of these hubs in more detail in the upcoming section.

Figure 8-3. *Synapse Studio hubs. Source: Microsoft*

Main Features of Synapse Studio

Synapse Studio provides many features, which are categorized or grouped into different hubs, as shown in Figure 8-3.

Synapse Studio also provides a notebook-based development experience, which is very much needed nowadays for any comprehensive data analytics platform. Nteract is an open source cloud-native notebook development framework, and it has been integrated by Microsoft into Synapse Studio. It provides an intuitive way of writing and running code to instantly visualize the data in the same space. That is a very important feature not just for real development work but also for ad hoc data analysis or, in data science terms, exploratory data analysis (EDA).

Home Hub

The first hub in Synapse Studio is the Home hub, which is immediately visible when you launch Synapse Studio. In the center of the Home hub, you will see the links to ingest, explore, analyze, and visualize your data, as shown in Figure 8-4. These are basically shortcuts to various tools available in Azure Synapse Analytics.

Figure 8-4. *Various links on Home hub in Azure Synapse Studio. Source: Microsoft*

Data Hub

The Data hub provides access to your Serverless Synapse SQL Pools as well as your Dedicated Synapse SQL Pools. It also gives access to external data sources and other linked services that you have created.

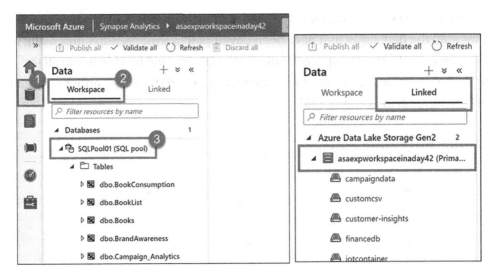

Figure 8-5. *Azure Synapse Studio, Data hub. Source: Microsoft*

The Data hub as shown in Figure 8-5 provides you with two tabs, which are as follows:

- The Workspace tab will give you a list of all the databases. You will also be able to see all the tables and stored procedures under those databases.

- The Linked tab will show you the storage account as well as linked services in your workspace.

Develop Hub

The Develop hub allows you to manage and develop SQL scripts, Synapse Spark notebooks, data flows, and Power BI reports. Therefore, the Develop hub is further divided into four different management subgroups, which include SQL scripts, notebooks, data flows, and Power BI. This is the hub where the real development work will happen, and all data roles, including data engineers, data analysts, data scientists, and so on, will access this hub for writing their code in it.

Under SQL scripts, you will be able to see the T-SQL scripts that you have published to your workspace. Within those SQL scripts, you will be able to execute commands against any of the Provisioned Synapse SQL Pools or Serverless Synapse SQL Pools. Under Notebooks, you will be able to see all the Synapse Spark notebooks that would have been executed using Synapse Spark Pool or Apache Spark Pool. Under Data flows, it will show all the data flows that have been developed in Synapse Pipelines using your Azure Synapse Analytics workspace. Under Power BI, you will be able to see all the Power BI reports embedded in your workspace. Figure 8-6 shows the options available for the Develop hub in Synapse Studio.

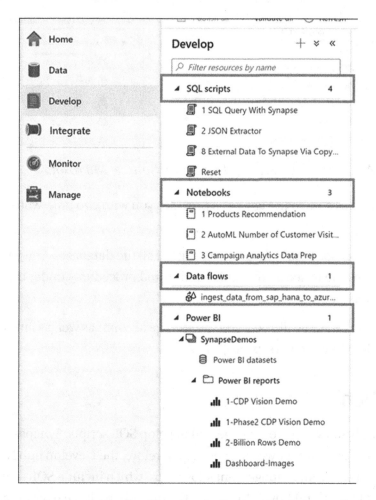

Figure 8-6. Azure Synapse Studio, Develop hub. Source: Microsoft

Integrate Hub

The Integrate hub is basically given to take you to the Synapse Pipelines interface. The look and feel are similar to Azure Data Factory, as Synapse Pipelines is architecturally similar to Azure Data Factory. This is an important hub from an integration point of view as you need not go to Azure Data Factory for any of your data integration and orchestration needs; the Integrate hub provides all those options within Synapse Studio itself.

Figure 8-7. *Azure Synapse Studio, Integrate hub. Source: Microsoft*

The Integrate hub as shown in Figure 8-7 provides a list of all the pipelines present in your Synapse workspace. It will also allow you to drag and drop various activities from the activities pane to the central pipeline development canvas. We have devoted a whole chapter to Synapse Pipelines in this book, so you can refer to that Chapter 7 for more input on what you can do with the Integrate hub from Synapse Studio.

Monitor Hub

The Monitor hub is meant to give you various options to monitor your pipeline runs, view the status of IRs, view Synapse Spark jobs, and more, along with historical details about activities that happened in your workspace. The Monitor hub is divided into two management groups, Integration and Activities, as shown in Figure 8-8. This is a good place to visit frequently when you are troubleshooting something, or if you want insights into resources being used in Azure Synapse Analytics.

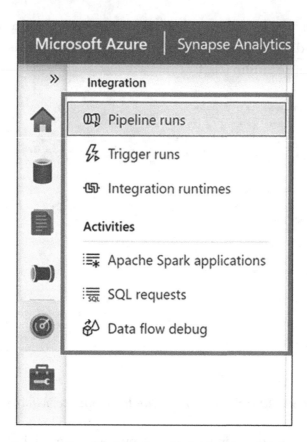

Figure 8-8. *Azure Synapse Studio, Monitor hub. Source: Microsoft*

Integration

In the Integration group, you have three different management subgroups, which include Pipeline Runs, Trigger Runs, and Integration Runtimes. Under Pipeline Runs, you can view all the pipeline run activities. It also allows you to view the run details with inputs and outputs for each activity in the data pipelines that have been executed or are being executed. It will also show any error messages that any data pipelines have generated in your workspace. If required, you can also stop a specific pipeline from here.

Under Trigger Runs, you can view all the pipeline runs initiated by automatic triggers. From here, you can also create recurring triggers, event-based triggers, as well as tumbling window triggers. Under Integration Runtimes, it will show a list of all the Self-Hosted Integration Runtimes (SHIR) and Azure Integration Runtimes along with their statuses.

Activities

In Activities, you have three different management sub-groups: Apache Spark Applications, SQL Requests, and Data Flow Debug. Under Apache Spark Applications, it shows all the Apache Spark applications that are currently running or have already run in your workspace. Under SQL Requests, you will be able to view all the SQL scripts executed directly by you or any other users or any other way, like from a Synapse Pipelines run. Under Data Flow Debug, it will show you all the active and previous debug sessions on the screen. While authoring a data flow, you have the option to enable the debugger and execute the data flow without needing to add it to a pipeline and trigger an execute to speed up the development activities.

Manage Hub

The Manage hub provides options to manage Synapse SQL pools, Synapse Spark pools, linked services, triggers, integration runtimes, access control, and private endpoints.

The Manage hub in Azure Synapse Studio is divided into four main management groups, as shown in Figure 8-9. These groups are Analytics Pools, External Connections, Integration, and Security. Each management group has one or more components under it.

Analytics Pools

The Analytics Pools group allows you to manage SQL pools as well as Apache Spark pools or Synapse Spark pools. Under SQL Pools, it will list all the Dedicated or Provisioned Synapse SQL Pools as well as Serverless or On-Demand Synapse SQL Pools. It allows you to add new pools as well as to manage existing pools. It provides options to either pause or scale SQL pools. If your SQL pool is not being used, then it is best practice to pause it to save costs.

For Apache Spark pools or Synapse Spark pools, it allows you to configure automatic scaling and automatic pause settings for the existing pools shown on the screen. It also allows you to provision a new Apache Spark Pool or Synapse Spark pool.

External Connections

Under the External Connections group, you have just one option, which is related to linked services. You can manage connections to many external resources via this option. It allows you to add linked services for Azure Data Lake Storage, Azure Key Vault, Power BI, and Azure Synapse Analytics. It also allows you to add new linked services.

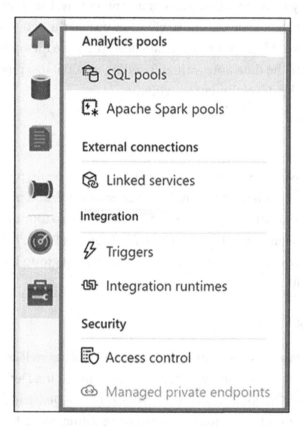

Figure 8-9. *Azure Synapse Studio, Manage hub. Source: Microsoft*

Integration

In the Integration group, you have two different options: Triggers and Integration Runtimes. It is the central location from which you can add or remove triggers for any of your Synapse Pipelines. Alternatively, there is an option to add triggers from Synapse Pipeline's canvas screen as well. Under Integration Runtimes, it shows the list of all the integration runtimes available in your Azure Synapse Analytics workspace.

Integration runtimes provide a compute environment for executing your Synapse Pipelines. If you hover over a specific integration runtime, then it will show you links for monitoring, code, and delete operations.

Security

For the Security group, you have two management options related to security: Access Control and Managed Private Endpoints. Access Control is the place where you will be able to add or remove users from one of three security groups, which include workspace admin, SQL admin, and Apache Spark admin. Managed Private Endpoints is the place from which you will be able to manage your private endpoints. It allows you to use a private IP address from within an Azure virtual network to connect to an Azure service or your own private link service. It shows the list of private endpoints, using which you will be able to connect to Serverless Synapse SQL Pools, Dedicated Synapse SQL Pools, and various development SDKs.

Synapse Studio Capabilities

We discussed the main features of Synapse Studio in detail in the previous section. That should have given you a fair idea about what tasks and actions you can perform using which hub in Synapse Studio. Those features are responsible for providing different capabilities within Synapse Studio. Let us discuss some of the important Synapse Studio Capabilities.

Data Preparation

Azure Synapse Analytics provides highly scalable and enterprise-grade data integration, data orchestration, and data transformation capabilities as part of Synapse Pipelines, which is nothing but a tightly integrated Azure Data Factory facility available within Synapse Studio. Data engineers require capabilities to design, build, and maintain robust data pipelines to bring in data from various disparate data sources. This is the prerequisite for data preparation tasks, as you will need the data to be ingested first to prepare it for your end goals. Synapse Studio also provides capabilities to transform the data via multiple options, including Synapse Spark Jobs and SQL Stored Procedures.

Additionally, Synapse Studio provides the capability to apply transformations to your data as part of your data preparation procedure using fully GUI-based data flows, which do not require any code to be written.

Data Management

The practice of collecting, storing, accessing, and using data securely and cost-efficiently is known as data management. Synapse Studio provides various data management capabilities. The Data hub within Synapse Studio is meant to provide data management capabilities. The Integrate hub allows you to collect the data. Azure Data Lake Storage Gen2, which is accessible via the Data hub, provides you with the facility to store, access, and use the data securely and cost-efficiently. You can easily access your databases from the Data hub within Synapse Studio as well.

Data Exploration

Data analysts, data scientists, and data engineers would like to explore data on an ad hoc basis before they finalize their respective workloads. So be it ad hoc data exploration or exploratory data analysis (EDA), both capabilities are present in Synapse Studio. You can easily explore the data using a Serverless Synapse SQL Pool, which is also known as an On-Demand Synapse SQL Pool. This gets provisioned along with your Synapse workspace and is immediately available to you for data exploration purposes. It is serverless and available on-demand, and you are charged based on the size of the data being processed; hence, it is cost effective as well for data exploration purposes. If required, you also have the option to write data exploratory Spark notebooks, as Synapse Studio provides Synapse Spark pools for this purpose.

Data Warehousing

When it was launched, Microsoft rebranded Azure SQL Data Warehouse as Azure Synapse Analytics. That means that inherently the latter supports all the enterprise-grade data warehousing capabilities that were present in the former. From Synapse Studio, you can easily create, populate, manage, and query your data warehouse. The well-known and proven massively parallel processing (MPP) engine is available to you as either a Dedicated or a Provisioned Synapse SQL Pool. Generally, you will have to ingest data into your data warehouse using ETL data pipelines, which you can do using Synapse Pipelines. You can also use SQL Store procedures or Synapse Spark notebooks to transform, aggregate, and consolidate your data for your data warehousing requirements. Through Synapse Studio, you can avail yourself of all these data warehousing capabilities.

Data Visualization

One of the key factors in being able to gain insights into your data is data visualization. It makes it easy for humans to comprehend any size of data, as data visualization makes it easy to detect any patterns, trends, and outliers in your data. Synapse Studio provides various built-in options to visualize your data. It provides some options as part of your Synapse Spark notebooks, which allows you to visualize your data using charts and graphs. Synapse Studio also allows you to use popular open-source data visualization-related libraries. The biggest capability for data visualization is of course the tight integration Synapse Studio provides with the Power BI workspace. Once you integrate your Power BI workspace into Synapse Studio, you can simply do everything from Synapse Studio that you can do from your Power BI workspace. Power BI is one of the most widely used data visualization tools in the industry, and most of its capabilities are directly available inside Synapse Studio.

Machine Learning

Generally, machine learning follows a pre-defined workflow, which includes business understanding, data acquisition, modeling, deployment, and scoring. Data pipelines help to acquire and ingest data for machine learning, something for which Synapse Pipelines capabilities can easily be leveraged. Ingested data will be prepared and explored further, along with visualization, and Synapse Studio provides all these capabilities, as we have highlighted in the preceding sections. For training machine learning models, we can use a Synapse Studio–based Spark notebook, which can be written using PySpark, Scala, or .NET, as all these languages are supported within Synapse Studio in Nteract-based notebook development experiences. These notebooks can be executed on Synapse Spark Pools for training models. You can use Spark MLib as well as Azure Automated Machine

Learning to train your models. For scoring models, Synapse SQL Pools support the T-SQL PREDICT function. Alternatively, Apache Spark Pool or Synapse Spark Pool can also be used to score the machine learning models using Synapse Studio.

Power BI in Synapse Studio

Azure Synapse Analytics allows you to integrate your Power BI workspace with it, as shown in Figure 8-10. Within Synapse Studio, you have an option to link your Power BI workspace through linked service creation. This allows you to create new Power BI reports and datasets from within Synapse Studio. This makes data visualization capabilities available within Synapse Studio, which makes Synapse Studio a comprehensive and complete data analytics tool.

Figure 8-10.   *Power BI in Synapse Studio. Source: Microsoft*

With its tight integration with Power BI, Synapse Studio brings multiple benefits to the table. It works as a single source of truth for Power BI reports, as well as centralized security. This also increases team collaboration as data analysts, data engineers, data scientists, and Power BI developers can use Synapse Studio as a tool for their respective workload developments. Therefore, sharing data and its output among teams becomes a lot easier. Data preparation is an important prerequisite for most of the Power BI reports, and Synapse Studio provides data preparation capabilities itself. The integration between Power BI and Synapse Studio also supports paginated reports as well as DirectQuery mode at scale. It is also possible to develop data flows using Power BI integrated into Synapse Studio.

As Azure Synapse Analytics supports Azure Machine Learning integration, this can be easily leveraged to develop AI-based predictive analytics reports using Power BI. This enables Power BI developers to utilize predictive models created by data scientists or to go with Power BI's own self-service model-creation feature. Power BI developers can either build a new data flow or reuse an existing data flow to train the predictive model, which can be further used to add prediction capabilities to the new incoming data. Thus, Synapse Studio's tight integration with Power BI does not just provide data visualization capabilities, but also provides machine learning capabilities.

How-To's

We have gone through some of the important components and features of the Azure Synapse workspace and Azure Synapse Studio in the previous sections. However, we have not looked at how to carry out a specific activity or task. Therefore, in this section we are going to discuss how you can perform a few of the important tasks for or using an Azure Synapse workspace and Azure Synapse Studio. Let us see how to create or provision a new Azure Synapse workspace first.

How to Create or Provision a New Azure Synapse Analytics Workspace Using Azure Portal

1. Log in to your Azure subscription by opening Azure Portal. Search for Azure Synapse Analytics, which you will find under Services. Click on Add to create an Azure Synapse Analytics workspace.

2. You will be presented with a wizard that has four different tabs, as shown in Figure 8-11. On the Basics tab, under the Project Details head, you will have to fill in the following details:

    a. Select the subscription in which you want to create a new Azure Synapse Analytics workspace.

b. Once you select the subscription, you will have to select an appropriate resource group from your subscription in which you want to create the new Azure Synapse Analytics workspace. If required, you can also create a new resource group from this screen, which can be used by your Azure Synapse Analytics workspace.

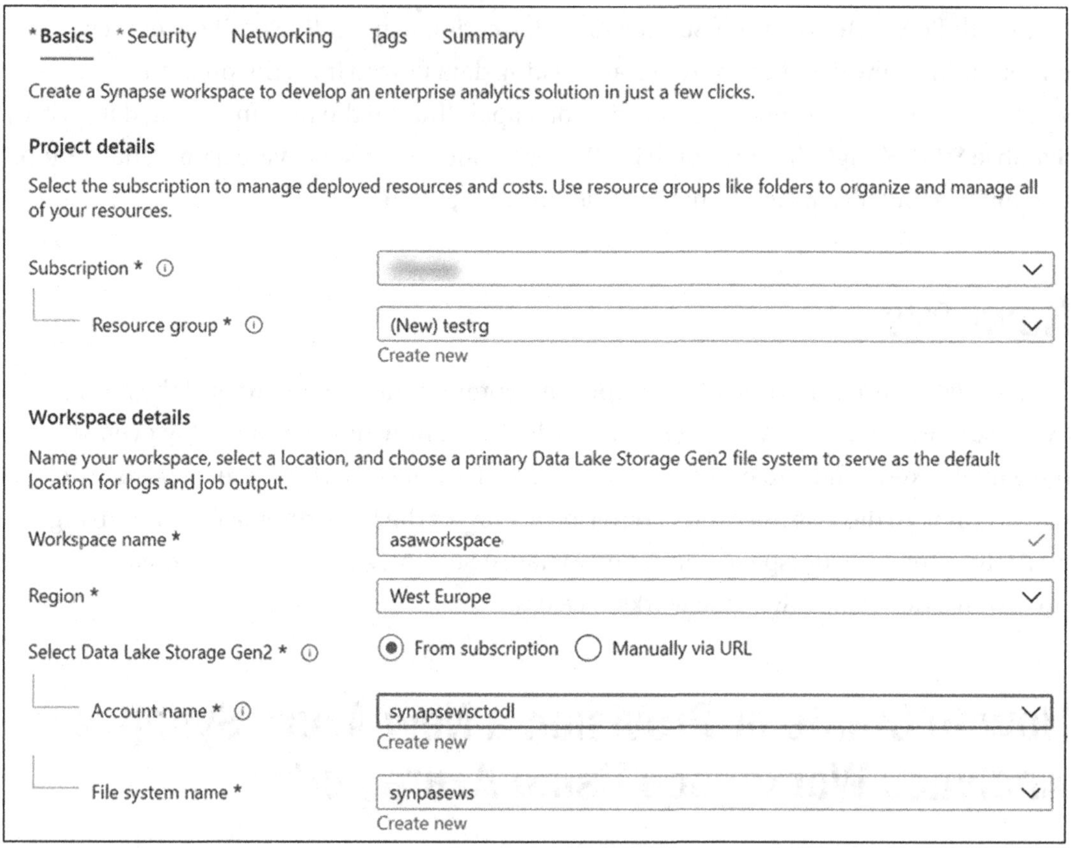

Figure 8-11. *Azure Synapse Analytics workspace creation wizard. Source: Microsoft*

3. On the Basics tab, under the Workspace Details head, you will have to fill in the following details:

a. For Workspace Name, you will have to pick any globally unique name.

b. For Region, select any region from the dropdown. Your new Azure Synapse Analytics workspace will be created in whatever region you select here.

c. Next, you will have to select an Azure Data Lake Storage (ADLS) Gen2, which will be served as the default location for storing logs and job outputs for Azure Synapse Analytics. Once you create an account for ADLS Gen2, you will have to select an existing file system name from the dropdown, or you can create a new file system name as well.

d. You will have to check the "Assign myself the Storage Blob Data Contributor role on the Data Lake Storage Gen2 account" box. This will allow the Azure Synapse Analytics workspace to use the ADLS Gen2 storage account for storing logs, job outputs, and Spark tables.

4. Go to the Review + Create tab and on that click the Create button. After a few minutes, your Azure Synapse Analytics workspace should be ready for use.

How to Launch Azure Synapse Studio

As we know, Azure Synapse Studio is the central place from which one can carry out most of the actions in Azure Synapse Analytics. So, once you have created or provisioned your first Azure Synapse Analytics workspace, the next thing you would like to do is to launch Azure Synapse Studio. There are a couple of ways in which you can do so. Let us look at the following options:

1. Log in to your Azure subscription via Azure Portal.

2. Search for Synapse Analytics Workspace using top search box.

3. Locate and open your Synapse Analytics workspace.

4. In the Overview section, go to Open Synapse Studio box as highlighted in Figure 8-12.

5. Click on Open button to launch Azure Synapse Studio.

Figure 8-12. *Open Synapse Studio from Azure Synapse Analytics workspace. Source: Microsoft*

There is also another option—to open Azure Synapse Studio directly. For that you can directly go to your web browser and use the following URL to open Azure Synapse Studio: `https://web.azuresynapse.net`. This will ask you to log in to your Azure Synapse Analytics workspace and after that it will take you directly to Azure Synapse Studio.

How to Link Power BI with Azure Synapse Studio

We know that we can link Power BI with Azure Synapse Studio. We also discussed that due to the very tight integration between Power BI and Synapse Studio, there are many benefits. Now, let us look at how to link a Power BI workspace with Synapse Studio.

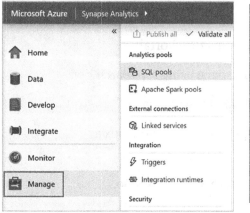
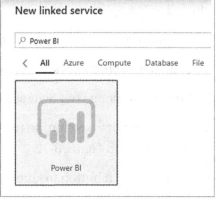

Figure 8-13. *Power BI as linked service in Synapse Studio. Source: Microsoft*

Let us look at the step-by-step process of linking a Power BI workspace with Azure Synapse Studio, as follows:

1. As prerequisites, you need to ensure that you have your Azure Synapse Analytics workspace and associated Azure Data Lake Storage (ADLS) Gen2 account already created for you. You also need to ensure that you have either a Power BI Professional or Power BI Premium workspace already created for you.

2. Launch Azure Synapse Studio from your Azure Synapse Analytics workspace.

3. On Synapse Studio, click on the Manage hub.

4. Under External Connections, click on Linked Services.

5. Click on the + New button, which will open a screen to create a new linked service.

6. Search for Power BI, and once it appears on the search results, click Power BI and click the Continue button on the screen, as shown in Figure 8-13.

7. On the New Linked Service (Power BI) screen, you will have to enter the following details as per Figure 8-14:

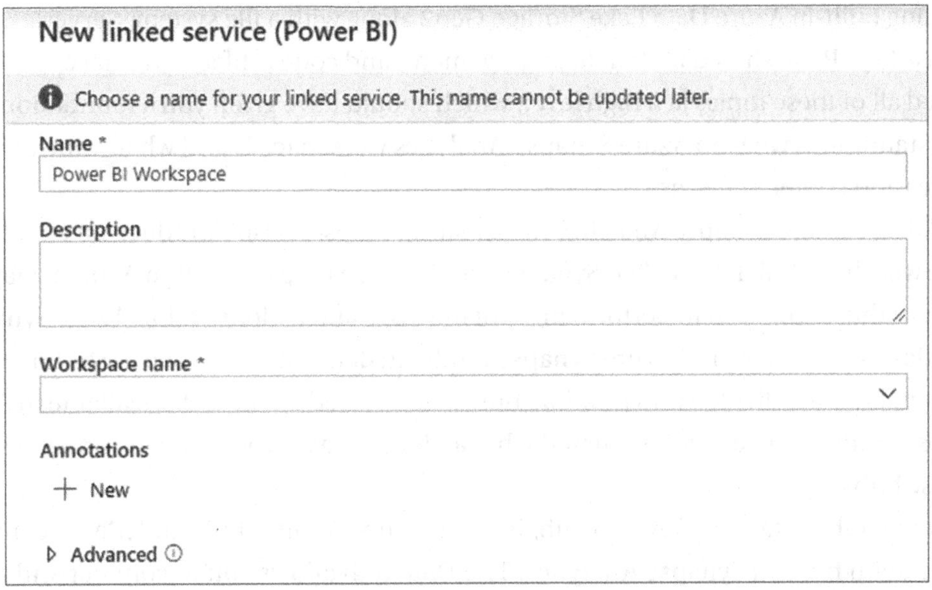

New linked service (Power BI)

ℹ️ Choose a name for your linked service. This name cannot be updated later.

Name *

> Power BI Workspace

Description

Workspace name *

> ⌄

Annotations

+ New

▷ Advanced ⓘ

Figure 8-14. Power BI, new linked service. Source: Microsoft

    a. **Name:** Provide a meaningful name for the linked service. It is a mandatory field.

    b. **Description:** Provide a description for your linked service. It is an optional field.

    c. **Workspace name:** Select a workspace name from the dropdown list. It is a mandatory field.

8. Click the Create button to create the linked service, which will link your Power BI workspace with Azure Synapse Studio.

Summary

We have looked at the Azure Synapse Analytics workspace, which was introduced soon after Microsoft rebranded Azure SQL Data Warehouse to Azure Synapse Analytics. Along with the Azure Synapse Analytics workspace, there are many features and components integrated seamlessly into Azure Synapse Analytics. Its workspace is the central place from which you can access and configure the various tools and technologies offered in Azure Synapse Analytics. It provides a space for close collaboration among your various IT teams, including data engineering, data analysis, data science, and data visualization.

The Azure Synapse Analytics workspace comes with many features and components, including built-in Azure Data Lake Storage Gen2 along with a file system, Serverless Synapse SQL Pool, shared metadata management, and code artifacts. We have covered all of these topics at a high level, which should have given you a foundational understanding of what an Azure Synapse Analytics workspace is and what some of its features and components are.

With the Azure Synapse Analytics workspace, Microsoft also introduced a cloud-native web-based GUI tool called Synapse Studio. It is a core tool within Azure Synapse Analytics that allows you to perform most of the tasks and actions within Azure Synapse Analytics. We have covered Azure Synapse Studio in detail. Azure Synapse Studio functionalities are divided into multiple hubs. We covered all the hubs available in Azure Synapse Studio in detail and discussed what actions or tasks you can perform using each of those hubs.

Home hub, Data hub, Develop hub, Integrate hub, Monitor hub, and Manage hub are the main hubs in Synapse Analytics. The Data hub allows you to connect and see the databases and storage account, which you can directly access from Synapse Studio.

The Develop hub provides you different options to develop SQL scripts, Spark notebooks, data flows, and Power BI artifacts. The Integrate hub is your gateway to Synapse Pipelines, which we discussed in detail in the previous chapter. The Monitor hub allows you to monitor pipeline runs, trigger runs, integration runtimes, Spark applications, SQL requests, and data flow debugs. The Manage hub provides the facility to manage Synapse SQL Pools, Synapse Spark Pools, linked services, triggers, integration runtimes, access control, and private endpoints from its interface.

We also looked at Power BI and Synapse Studio's tight integration and the benefits we get out of this integration. Having Power BI inside Synapse Studio is definitely amazing since it adds the capability to generate all those funky-looking Power BI reports directly from Synapse Studio. It also provides you the capability to generate paginated reports, create datasets, make data flows, and have AI capabilities by allowing it to predict values for incoming data.

Toward the end of the chapter, we discussed how to carry out certain important activities for an Azure Synapse Analytics workspace and Azure Synapse Studio. We looked at how to create or provision a new Azure Synapse Analytics workspace in your Azure subscription. We also looked at how to launch Synapse Studio once you have provisioned or created an Azure Synapse Analytics workspace. Synapse Studio can also be integrated with Power BI, and we covered that as well toward the very end of the chapter.

I am hoping that this chapter has given you a solid foundation on the Azure Synapse Analytics workspace and Azure Synapse Studio, as we have discussed their features, components, capabilities, and more. In the next chapter, we are going to look at Azure Synapse Link, which is one of the most advanced topics in Azure Synapse Analytics. So, see you in the next chapter!

CHAPTER 9

Synapse Link

For any traditional data analytics project, be it a data warehouse or a data lake or a data lakehouse implementation, you will have to ingest the data from disparate source systems. Historically, source systems are business applications being used continuously to carry out various business operations. These source systems generate and store a large amount of data in various formats. If you try to generate business intelligence while these source systems are in use for business purposes, it may create bottlenecks in your business transaction processing. No business wants to see that type of impact on their core business application.

To solve this puzzle, data from source systems are ingested regularly at a set frequency using batch ingestion to a different system; that system is where the data analytics workload will be carried out. It takes time to ingest data from a source system to a data analytics platform, as you have to build data pipelines and then run and monitor them on a schedule to ingest the data to a data warehouse, data lake, or data lakehouse. There is a significant delay in this process as you have to wait for your data pipelines to finish before you can generate any business intelligence or insights from your data.

Nowadays, in addition to batch ingestion, there are new tools and technologies that support the live-streaming of data into your data analytics platform. This is called stream ingestion or stream processing or data streaming. This happens in near real time or real time depending on how you integrate your source systems with your data analytics platform. Here too, you have the source system and target system on different platforms. But what if there were some technology or tool that did not require you to build batch ingestion or stream ingestion data pipelines and could still run the data analytics workload instantly without impacting business transactions being carried out in the source systems? Well, that is exactly what Synapse Link does for you. It is a new feature in public preview at the time of writing this book. This may be one of the most advanced and exciting features of Azure Synapse Analytics yet. In this chapter, we are going to look at Synapse Link in much more detail. Apart from introducing and discussing what Synapse Link is, we will also discuss its benefits, features, and use cases. Before that, we will start with some basic concepts like OLTP, OLAP, and HTAP.

201

© Bhadresh Shiyal 2021
B. Shiyal, *Beginning Azure Synapse Analytics*, https://doi.org/10.1007/978-1-4842-7061-5_9

OLTP vs. OLAP

Any system that supports the processing of various business transactions online is generally known as an on-line transaction processing (OLTP) system. Generally, OLTP systems remain in use during business hours to support the various business operations. These systems typically support select, insert, update, and delete operations on the back-end database in real time. These systems are designed to take heavy concurrent load. That means that multiple users can connect with the front end of your OLTP system and carry out multiple different transactions, which will immediately get committed to your back-end database. OLTP systems are write-intensive as well read-intensive.

Any system that supports the simple to complex analysis of a large volume of data online is generally known as an on-line analytical processing (OLAP) system. Generally, OLAP systems are used by data scientists, data analysts, BI professionals, and knowledge workers. These systems typically support bulk loading of data as well as complex queries related to data analytics for a large volume of data. These systems are read-intensive, as most of the users would like to read or query the data stored in these systems, while the writing of data to these systems happens generally once a day, or a relatively smaller number of write operations will take place against these systems. They can support an organization's requirements around data analysis, data analytics, data science, and business intelligence. Transaction data from OLTP systems will get ingested into OLAP systems. Therefore, your OLAP system should support data ingestion from diverse data sources.

The OLAP system will receive data from OLTP systems and other systems that will be further processed by the OLAP systems to generate deeper insights into the data. There has to be a robust data ingestion process in place. These processes will take some time, and only once the full data is ingested you will be able to run your data analytics workload on the OLAP systems. Generally, ETL/ELT tools will be used to extract, transform, and load the data into the OLAP systems. These tools help to develop the data pipelines rapidly so that you can create a bridge between your OLTP and OLAP systems to ensure the smooth movement of your data. This will introduce some delay in getting insights from your data as the data ingestion process will take some time to complete on a daily basis. Another problem with this setup is that you will have to work with slightly older data. Typically, the concluding day's data will be loaded during the night, and the next morning you will be able to generate insights from it. So, this setup not only delays the insights from the data, but also provides insights on slightly stale data.

To overcome the drawbacks of the OLAP and OLTP setup, there is a relatively new concept called HTAP, which we are going to discuss in more detail in the next section.

What Is HTAP?

OLTP + OLAP = HTAP

HTAP stands for **h**ybrid **t**ransaction and **a**nalytical **p**rocessing. It is an advanced and rapidly emerging trend in the data analytics space. As its name suggests it combines OLTP and OLAP into one. By doing so, it breaks the barrier between the two and provides capabilities to generate instant insights from your data. In other words, HTAP can enable real-time analytics on your data without making any adverse impact on your transactional system. It also avoids any data synchronization issues that may emerge if you are ingesting data from OLTP to OLAP. This is because you do not need to implement any data ingestion pipelines. So, HTAP is a no-ETL data analytics concept.

Benefits of HTAP

HTAP has several benefits over traditional OLTP and OLAP systems. Let us discuss these benefits briefly.

No-ETL Analytics

In traditional OLTP and OLAP systems, it is required to set up a data ingestion pipeline through any of the ETL/ELT tools, as this architecture requires the movement of data from the transactional systems to the analytical systems. Building an ETL pipeline is definitely not an easy task. It requires a significant amount of time to design, build, deploy, and then maintain those data ingestion ETL pipelines. As the number of source systems and the size of the data keep on increasing, it will require continuous effort to build and maintain those data ingestion ETL pipelines to meet your data integration and data orchestration requirements. HTAP does not require you to build data ingestion pipelines, so it is a no-ETL data analytics architecture.

Instant Insights

In traditional OLTP and OLAP systems, you have to wait for your data ingestion pipelines to complete the ingestion process before you can run your data analytics workload on the ingested data. This is a required step since you have to ingest your transactional data from your OLTP systems to your OLAP systems for running your complex and long-running analytics queries. So, there will be a significant delay to generate insights from your operational data.

This is not the case with the HTAP system. HTAP can generate instant insights on your operational data without impacting the performance of your transactional systems. As there is no ETL required for data ingestion pipelines in HTAP, the operational data is made available in almost real time so you can generate instant insights into your data. This is one of the biggest driving factors in today's scenario as every business wants to generate meaningful and actionable insights from their data at the earliest possible time. This gives an edge to those organizations that are using HTAP, as their go-to-market time will be reduced heavily compared to those organizations that are not using it.

Reduced Data Duplication

In traditional architecture in which OLTP and OLAP systems are involved, multiple copies of the same data get copied to different systems. It also involves significant effort to monitor and synchronize these data copies to avoid data inconsistencies and other related issues. In HTAP, the need to create multiple copies of the same data can either be completely avoided or at least be reduced significantly. So, indirectly, it reduces the effort required to create and maintain multiple copies of the same data, as HTAP is capable of supporting your data analytics workload without creating copies of all your operational data.

Simplified Technical Architecture

In traditional architecture with OLTP and OLAP, you have to extract the data from the operational systems, apply transformations on it, and then load the transformed data into your analytical system. This type of architecture requires many components like ETL/ELT tools, an enterprise service bus, messaging tools, and event management tools. This increases the complexity of your architecture.

HTAP will simplify your technical architecture. You do not need all those tools in HTAP as you need not copy or replicate or ingest all the data from your operational systems to an analytical system. It will inherently support synchronization without all those tools and technologies. So, compared to OLTP- and OLAP-based architecture, HTAP will definitely simplify your technical architecture, and thus the implementation will also be much simpler and less time consuming.

What Is Azure Synapse Link?

So far, we have covered some fundamental concepts around OLTP, OLAP, and HTAP. We also discussed HTAP in a little more detail and went through some of its benefits. Now, it is time to link it back to Azure Synapse Analytics. Azure Synapse Link is a new feature in Azure Synapse Analytics that bridges the gap between OLTP and OLAP systems, and hence is an HTAP implementation within Azure Synapse Analytics.

In Microsoft's words, Azure Synapse Link is a cloud-native Hybrid Transactional and Analytical Processing (HTAP) capability that enables you to run near-real-time analytics over your operational data. Currently it supports Azure Cosmos DB only. That means that the Azure Synapse Link feature can be used only with Azure Cosmos DB as of now. This feature is in preview at the time of writing this book. Basically, Azure Synapse Link creates a tighter and more seamless integration between Azure Cosmos DB and Azure Synapse Analytics (Figure 9-1).

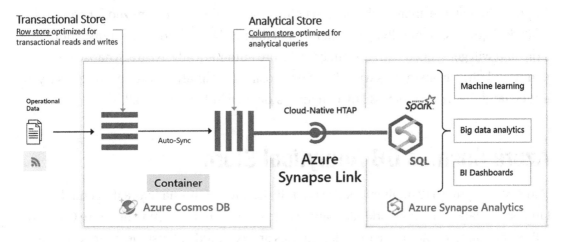

Figure 9-1. *Azure Synapse Link. Source: This image is taken from Microsoft Documentation titled "What is Azure Synapse Link for Azure Cosmos DB?" dated November 30, 2020.* `https://docs.microsoft.com/en-us/azure/cosmos-db/synapse-link?toc=/azure/synapse-analytics/`

It is better to understand Azure Cosmos DB first before trying to understand the architecture for Azure Synapse Link. So, let us briefly discuss Azure Cosmos DB and the two different types of stores—transaction store and analytical store—that are used in Azure Synapse Link. You can refer to Figure 9-1 to understand the architecture of Azure Synapse Link at a high level.

Azure Cosmos DB

Azure Cosmos DB is a fully managed and truly globally distributed no-SQL database service available on Azure. It gives you single-digit millisecond response times, automatic and instant scalability, guaranteed speed at any scale, and multiple API model options to choose for your workload. Apart from an obvious SQL or Core API, it also supports Cassandra API, MongoDB API, Gremlin API, and Table API. With Azure Cosmos DB, you need not worry about database administrator–related activities, as it is fully managed by Microsoft for you. This means that it supports automatic management, updates, and patching as well.

For Azure Synapse Link, you need to understand Azure Cosmos DB transactional store as well as Azure Cosmos DB analytical store. As their names suggest, they are meant for different purposes and have different capabilities as well. The transactional store is a row store, which is optimized for transactional read and write operations. Opposite to that, the analytical store is a column store, which is optimized for complex analytical queries. Generally, your operational data will be stored in the transaction store and will get synced with the analytical store automatically. Azure Synapse Link tightly and directly integrates with the Azure Cosmos DB analytical store. So, let us try to understand the Azure Cosmos DB analytical store in a little more detail.

Azure Cosmos DB Analytical Store

The Azure Cosmos DB analytical store is a fully isolated columnar store that enables large-scale data analytics workloads against the operational data in your Azure Cosmos DB. Since your complex data analytics queries are executed against the Azure Cosmos DB analytical store, it will not have any adverse impact on your operational data, which is stored in your Azure Cosmos DB transactional store. Since we are going to run analytical queries on the Azure Cosmos DB analytical store, it is schematized to provide optimized performance. However, the Azure Cosmos DB transactional store is schema-agnostic to support transactions from your business applications.

You may be wondering why we need to have a separate Azure Cosmos DB analytical store while our data is already stored in the Azure Cosmos DB transactional store. Is it not possible to run complex data analytics queries directly on the Azure Cosmos DB transactional store? Well, technically it is possible to do so, but in doing that we would face a number of different challenges. The first challenge is about the performance of your Azure Cosmos DB transactional store. You will get performance bottlenecks if you run your complex analytics queries directly on your operational data, as it is going to impact transactions being carried out by your business users. Your Azure Cosmos DB transaction store's primary responsibility is to serve business transactions and not analytical queries. Another challenge would be around the format in which the data is stored. For Azure Cosmos DB transactional stores, data is stored in rows to support optimum performance for transactional read and write operations. Now, if you run analytical queries on this store, your queries will not give you optimum performance, as it is a row store and not a column store, which is best suited for your complex analytical queries.

Therefore, it is highly recommended that your transactional or operational data store be different from your analytical data store. In a traditional setup, you have to ingest the data from your transactional data store to your analytical data store at a regular interval. Since these data ingestion pipelines are going to read data from your transactional or operational data store, it is also recommended that you run your data ingestion pipeline during off-peak hours and less frequently to minimize the performance impact on your transactional or operational data store.

To take care of all these issues, the Azure Cosmos DB analytical store is used in Azure Synapse Link. It avoids complexity and latency challenges that cannot be avoided with traditional ETL data ingestion pipelines. As per Azure Synapse Analytics' architecture, the Azure Cosmos DB analytical store can automatically synchronize operational or transactional data from the Azure Cosmos DB transactional store. Along with automatic synchronization, it will take care of converting the data from row-based storage to column-based storage in a Azure Cosmos DB analytical store, as a columnar format is more appropriate for complex and large-scale analytical queries.

Let us look at some of the important features of the Azure Cosmos DB analytical store at a high level.

Columnar Storage

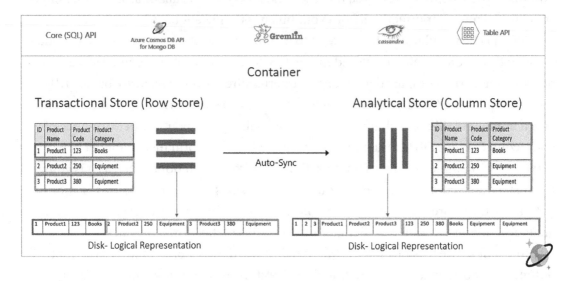

Figure 9-2. *Columnar storage for analytical workload. Source: This image is taken from Microsoft Documentation titled "What is Azure Cosmos DB Analytical Store?" dated March 6, 2021.* `https://docs.microsoft.com/en-us/azure/cosmos-db/analytical-store-introduction`

It is important to understand how columnar storage works internally for Azure Cosmos DB in the analytical store. If you look at Figure 9-2, you can easily see that in a transactional store, the data is stored in a row-oriented structure. Logically, all the fields of a row are stored in serialized format. However, once the data gets auto-synced to the Azure Cosmos DB analytical store, the same data is stored in a column-oriented structure. Logically, the same columns from each row are stored in a serialized format. A column store is best for optimum query performance, as it will be able to serialize the similar columns of data together, and that way it will result in an overall reduction in disk I/O. Reduced disk I/O will result in better overall query performance.

Decoupling of Operational Store

The Azure Cosmos DB analytical store allows us to decouple the operational store completely. Apart from the automatic data synchronization, there is no other impact on the Azure Cosmos DB transactional store. However, the automatic data synchronization process is executed internally in such a way that it will not have any performance impact on your Azure Cosmos DB transactional store. This means that the auto-sync process does not interfere with read and write operations happening on your Azure Cosmos DB transactional store. Complex and large-scale data analytics queries that may always run longer than a normal read or write operation will be executed against your data synced to the Azure Cosmos DB analytical store.

Automatic Data Synchronization

This is one of the most important features of the Azure Cosmos DB analytical store, and this is the feature due to which we have Azure Synapse Link available in Azure Synapse Analytics. It is a fully managed capability of Azure Cosmos DB in which all the inserts, updates, and deletes being executed in the Azure Cosmos DB transactional store are automatically synchronized to the Azure Cosmos DB analytical store. This auto-sync process happens in near real time. Generally, you will have latency numbers within two minutes. However, if you are using a shared throughput database that contains a large number of containers, then in that case the auto-sync latency of each of those containers could be slightly higher than two minutes, but it will still be within five minutes. This is an important limit to keep in mind while designing your solution so that you make a well-informed decision. This will also help you to justify any delay in updated data's appearing in the Azure Cosmos DB analytical store.

SQL API and MongoDB API

As we mentioned previously, Azure Cosmos DB supports multiple API models. It supports SQL or Core API, Mongo DB API, Cassandra API, Gremlin API, and Table API. Out of these five models, the Azure Cosmos DB analytical store is currently supported by Azure Cosmos DB SQL or Core API and Mongo DB API only. The remaining three API models are not supported, and hence those cannot be used with Azure Synapse Link for Cosmos DB.

Analytical TTL

TTL stands for "time-to-live." Analytical TTL indicates how long data should be retained in your analytical store for a container. If you have enabled an analytical store at the container level then each insert, update, and delete operation will automatically be synced from your transactional store to the analytical store irrespective of your

209

Transactional TTL setting. Thus, you can control the retention of your operational data in the analytical store by setting up the appropriate value for the Analytical TTL.

You can set up three different types of values for Analytical TTL. If you set "0" as the value or the value is missing and you have not set any value for it, no data will be replicated from your transactional store to your analytical store. If you set it to "-1" then it will retain all the historical operational data in your analytical store indefinitely. If you set "n," which is any positive number, as the Analytical TTL, then your operational data copied to the analytical store will be retained for "n" seconds after the "last modified" date in the transactional store. This type of value for Analytical TTL gives us the ability to decide how long we want to retain the data after it is last modified. You can set the Analytical TTL value from Azure Portal very easily. However, you can also use Azure SKD, CLI, or PowerShell scripts to set the various types of values for Analytical TTL.

Automatic Schema Updates

As we have discussed previously, Azure Cosmos DB transactional store does not have any schema while Azure Cosmos DB analytical store uses schematized data to support complex query processing. We have also discussed that Azure Cosmos DB transactional store have row-oriented storage while Azure Cosmos DB analytical store have columnar storage. We have also seen that there is an automatic data synchronization process which is responsible to sync the data from Azure Cosmos DB transactional store to Azure Cosmos DB analytical store. As part of these automatic sync process, it also ensures that it converts the data from schema-less row- oriented storage to schematized column-oriented storage.

Now, let us examine a scenario where there is a change in your transactional store data. You start to have additional fields as part of your transactional data store. What will happen when the data from the transactional store gets synced with the analytical store? Well, the auto-sync process will take care of schema updates automatically. That means that without any manual intervention, the schema updates will be reflected in the analytical store automatically.

Cost-Effective Archiving

The Azure Cosmos DB analytical store allows us to implement automatic tiering of data from the Azure Cosmos DB transactional store. It is important to note that the Azure Cosmos DB analytical store is optimized for storage cost, unlike the Azure Cosmos

DB transactional store. So, this means that you can store historical data into your Azure Cosmos DB analytical store very cost effectively. It can be used as cost-effective archival storage by storing data in the Azure Cosmos DB analytical store for longer-term retention.

You have the option to set different values for Transactional TTL and Analytical TTL. In order to take advantage of the cost-effective storage found in Azure Cosmos DB analytical stores, you can set a very low retention value for Transactional TTL and, opposite to that, set a very high retention value for the analytical store.

Scalability

The Azure Cosmos DB transactional store uses horizontal partitioning. This allows it to scale the storage and throughput elastically, and it does not require downtime during scaling operations. Horizontal partitioning provides scalability in the automatic synchronization process to ensure that the data is synced in near real time to the Azure Cosmos DB analytical store from the transactional store.

This scalability plays an important role in maintaining the automatic data synchronization irrespective of transactional traffic throughput. Let us assume that there are 5,000 operations per second. Automatically synchronizing data to the analytical store will not impact the throughput of your transactional store. This throughput will still not get impacted even if you have 1 million transactions per second that the automatic data synchronization process will have to sync with the analytical store from the transactional store.

When to Use Azure Synapse Link for Cosmos DB

Whenever there is a new or advanced feature available, it is always difficult to decide when to use that feature and which are the most suitable use cases. So, let us discuss the recommended use cases for Azure Synapse Link.

If you are already using Azure Cosmos DB for your project and you want to run data analytics, business intelligence, and machine learning workloads on your operational data, then in that case it makes sense to use Azure Synapse Link. It will give you a more integrated and comprehensive analytics platform without having any adverse impact on your operational data store. The best part is that you need not to run ETL processes to ingest the operational data into a separate analytics platform. You can take advantage of the benefits of an Azure Cosmos DB analytical store that gets synced with your Azure Cosmos DB transactional store automatically. This will also not impact the performance of your Azure Cosmos DB transactional store, which you may already be using to support your transaction system to meet your business requirements.

However, there are some scenarios in which Azure Synapse Link is not recommended. For example, if you have traditional data warehouse requirements in which you need to have high concurrency and you also need to manage your data warehouse workload, then in that scenario you should not look at Azure Synapse Link as a solution.

Azure Synapse Link Limitations

All technologies and tools have some limitations, and so too does Azure Synapse Link. It is important to know these limitations, as they will help you to make appropriate architectural decisions with respect to Azure Synapse Link. Let us look at the following limitations:

1. The biggest limitation is that currently only Azure Cosmos DB is supported in Azure Synapse Link. That means that if your operational data store is not already on Azure Cosmos DB then you will not be able to use Azure Synapse Link. However, it is expected that Microsoft will add additional Azure-based databases to the list of supported databases for Azure Synapse Link in the near future. Azure SQL Database, Azure Database for PostgreSQL, and Azure Database for MySQL may support Azure Synapse Link going forward.

2. Azure Cosmos DB supports multiple API models, which include SQL or Core API, Mongo DB API, Cassandra API, Gremlin API, and Table API. As mentioned previously, currently Azure Synapse Link supports only Azure Cosmos DB SQL or Core API and Azure Cosmos DB Mongo DB API. The remaining three API models are not supported as of now.

3. In Azure Synapse Analytics, you have three different compute engines available to you, which include Serverless Synapse SQL, Provisioned Synapse SQL, and Synapse Spark. Out of these three options, you cannot use Provisioned Synapse SQL to query the data stored in your Azure Cosmos DB analytical store.

4. Although you can enable Azure Synapse Link on your existing as well as new Azure Cosmos DB accounts, that is not possible to do for the existing containers in your Azure Cosmos DB. That means that you will have to migrate your data from existing containers to new containers to enable Azure Synapse Link. However, the good part is that Microsoft provides some options to migrate your data from existing containers to new containers. The following link contains details about those options: `https://docs.microsoft.com/en-us/azure/cosmos-db/cosmosdb-migrationchoices`

5. From an Azure networking point of view, there is no option to isolate the Azure Cosmos DB analytical store in Azure Synapse Analytics using managed private endpoints.

6. Another minor but noteworthy limitation is that there is some latency in the automatic synchronization process that is responsible for copying the data from the Azure Cosmos DB transactional store to the Azure Cosmos DB analytical store. Even though the latency is within two to five minutes, it may be a limitation for some of the use cases in which you want real-time synchronization or real-time access to your data. As Azure Synapse Link can provide only near-real-time synchronization, it may not be an ideal choice where you need real-time data only.

7. Currently there is no support for backup and restore for a Azure Cosmos DB analytical store. However, you can back up and restore your Azure Cosmos DB transactional store easily. Even though you are able to back up a transactional store that you have enabled for Synapse Link, when you restore it, it will only restore your transaction store containers and not your analytical store containers.

Azure Synapse Link Use Cases

There are a number of different industry use cases in which we can think of applying features from Azure Synapse Link. You can think of industry scenarios in which there is a need to generate near-real-time analytics on operational data. So, let us try to understand some of the industry use cases for Azure Synapse Link.

Industrial IOT

There are a ton of innovations happening in the industrial IoT space. Industrial IoT device sensors generate a lot of data that can be put into use to carry out predictive maintenance of crucial industrial machinery to avoid sudden downtime that can result in a huge loss for the organization. There are many use cases in which Azure Synapse Link can be used to predict the maintenance schedule for such machineries.

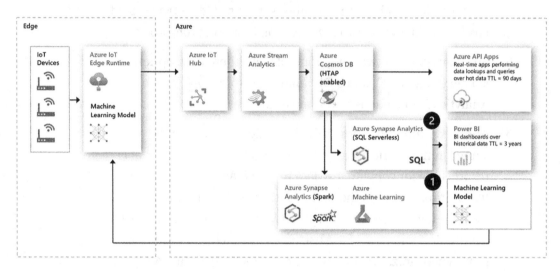

Figure 9-3. *Predictive maintenance using industrial IoT through Azure Synapse Link. Source: This image is taken from Microsoft Documentation titled "Azure Synapse Link for Azure Cosmos DB: Near real-time analytics use cases" in a section named "IOT Predictive Maintenance" dated May 19, 2020.* `https://docs.microsoft.com/en-us/azure/cosmos-db/synapse-link-use-cases#iot-predictive-maintenance`

Let us try to understand Figure 9-3 so we know how the overall architecture looks and what various data or process flows are shown in the diagram.

On the extreme left side of the diagram, you can see the Edge components, which include industrial IoT devices and Azure IoT Edge Runtime. Numerous industrial IoT devices will continuously generate a huge amount of sensor data, which will be fed to Azure IoT Edge Runtime for further processing. Azure IoT Edge Runtime will ingest the data into Azure IoT Hub, from which the data will be utilized for stream processing through Azure Stream Analytics. Here, Azure Stream Analytics is responsible for pushing the data to HTAP-enabled Azure Cosmos DB. As it is HTAP enabled, it will be able to use Azure Synapse Link features available in Azure Synapse Analytics.

There are many different use cases that can be implemented by using Azure Synapse Link with industrial IoT devices. Let us understand two of those use cases shown in the preceding diagram.

Predictive Maintenance Pipeline

The Azure Synapse Link can store the operational data from those industrial IoT devices in the Azure Cosmos DB analytical store. Due to the automatic synchronization of data, the current and historical operational data available in the Azure Cosmos DB analytical store can be used directly by Azure Synapse Analytics. Here, Synapse Spark, which is Microsoft's own implementation of the open source Apache Spark engine, is being used to read and prepare the data and feed it to the Azure Machine Learning service. Azure Machine Learning will train the anomaly detector model, and the trained model will be deployed back to Azure IoT Edge Runtime. This creates a loop in which continuous retraining of the predictive maintenance model based on the anomaly detector is done.

In this use case, the real benefit comes in terms of the continuous retraining and deployment of the latest model to Azure IoT Edge Runtime. This keeps your machine learning model fresh as it is trained by the latest data available from the Azure Cosmos DB analytical store. This allows you to accurately predict the maintenance schedule for crucial industrial machinery in near real time. It avoids the unscheduled breakdown of that machinery, which can save a lot of effort and money.

Operational Reporting

As we discussed in the previous section, industrial IoT devices generate a vast amount of sensor data, which gets accumulated in the Azure Cosmos DB analytical store. As the Azure Cosmos DB analytical store can also store a large volume of historical data along with current data, an opportunity is created to know the historical trends from sensor data. Here, a Serverless Synapse SQL Pool can be used to read and consolidate the data from the Azure Cosmos DB analytical store. Later, the consolidated data can be used to create a Power BI dashboard. Operational reporting via this flow provides greater and deeper insights into the vast amount of sensor data collected from those numerous industrial IoT devices.

Due to the rise of Digital Twins architecture, industrial organizations are increasingly creating digital copies of their machines. Digital Twins are virtual replicas of physical devices. These digital copies help them to meet their operational reporting needs instantly, without much delay.

Real-Time Applications

There is a third use case that is not directly related to Azure Synapse Analytics but is important to know since it also relies on HTAP-enabled Azure Cosmos DB. We can develop real-time applications using the Azure API App, which will fetch the hot data from the Azure Cosmos DB analytical store using various data lookups and queries. This can easily be supported by the Azure Cosmos DB analytical store since it can sync the data from the Azure Cosmos DB transactional store with a latency of two to five minutes; hence, it can give near-real-time data to the consuming real-time applications. These applications provide almost-real-time updates about those machines, which allows the monitoring team to take corrective measures at the right time on the right machine either automatically or manually.

Real-Time Personalization for E-Commerce Users

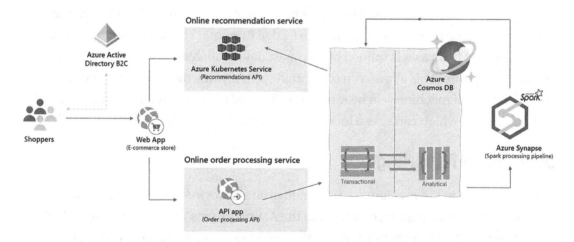

Figure 9-4. *Real-time personalization in e-commerce. Source: This image is taken from Microsoft Documentation titled "Azure Synapse Link for Azure Cosmos DB: Near real-time analytics use cases" in a section named "Real-time Personalization" dated May 19, 2020.* https://docs.microsoft.com/en-us/azure/cosmos-db/synapse-link-use-cases#real-time-personalization

E-commerce portals being used by retailers compete to provide the most engaging and personalized experience to their users and businesses. Azure Synapse Link can be leveraged to provide personalized experiences along with customized products and services, which may be highly personalized for each of those users and businesses. Let us try to understand the process flow, which is depicted in Figure 9-4.

The online e-commerce portal is shown to use two different APIs to create personalized experiences, including customized products and services. The Order Processing API uses the Azure API App to store the order and its related details into the Azure Cosmos DB transactional store. Here, Azure Cosmos DB is HTAP enabled, which means that Azure Cosmos DB transactional store data will be automatically synced to the Azure Cosmos DB analytical store. We can use a Synapse Spark Pool to process the near-real-time data available from the Azure Cosmos DB analytical store to train the machine learning model, which is used by the Recommendation API deployed in Azure Kubernetes Service (AKS). As the Recommendation API is fed with the latest available machine learning model by Synapse Spark, it will be able to generate real-time personalized experiences for users and businesses.

How-To's

Let us look at how to carry out some of the important activities involving Azure Synapse Link.

How to Enable Azure Synapse Link for Azure Cosmos DB

By default, Azure Cosmos DB accounts do not get provisioned with Azure Synapse Link. You can enable Azure Synapse Link from the Azure Portal once it is provisioned in your subscription. Let us check the step-by-step process, which is as follows:

1. Sign in to the Azure Portal and get into the subscription in which you want to enable Azure Synapse Link for Azure Cosmos DB.

2. Navigate to your existing Azure Cosmos DB account or create a new one, if required.

3. Under Settings, click on Features, which will open the features pane for you.

4. Select Synapse Link from the features list, as shown in Figure 9-5.

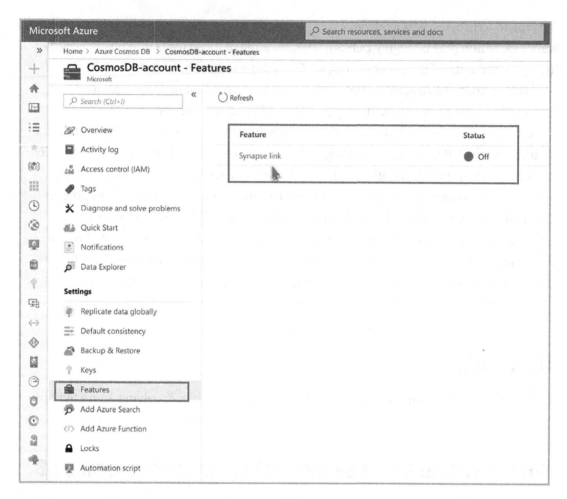

Figure 9-5. *Select Synapse Link feature. Source: Microsoft*

5. Next, it will prompt you to enable the Synapse Link on your
    Azure Cosmos DB account. Click the Enable button as shown
    in Figure 9-6 and wait around one to five minutes to allow the
    background process to finish. After that, your Azure Cosmos DB
    account will be enabled for Azure Synapse Link.

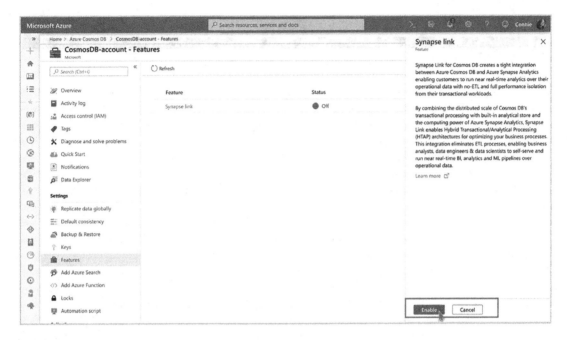

Figure 9-6. *Enable Synapse Link. Source: Microsoft*

How to Create an Azure Cosmos DB Container with Analytical Store Using Azure Portal

One important point to note here is that once you enable Azure Synapse Analytics on your Azure Cosmos DB account, it is not going to automatically create a Cosmos DB container with an analytical store by default. That means that once you enable the Azure Synapse Link for your Azure Cosmos DB account, you will have to take the following steps to create an Azure Cosmos DB container with an analytical store through Azure Portal:

1. Sign in to the Azure Portal and navigate to your Azure Cosmos DB account.

2. Open the Data Explorer tab and select New Container. Enter a name for your database, container, partition key, and throughput details. Ensure you turn on the analytical store option as well.

Figure 9-7. *Turn on analytical store. Source: Microsoft*

3. If you have not enabled the Synapse Link on your Azure Cosmos
    DB account then it is going to prompt you to do so because that
    is a prerequisite to create an Azure Cosmos DB container with
    an analytical store enabled. If prompted to enable Synapse Link
    then do so; you will have to wait for around one to five minutes to
    ensure that it gets enabled successfully.

4. Click the OK button as shown in Figure 9-7 to create an Azure
    Cosmos DB container with an analytical store enabled for it.

How to Connect to Azure Synapse Link for Azure Cosmos DB Using Azure Portal

To connect to Azure Synapse Link for Azure Cosmos DB, we will use Azure Synapse
Studio. It is assumed that you have already enabled Synapse Link on your Azure Cosmos
DB account. If that is not the case, then please go through the steps mentioned earlier
before moving ahead, as that is the prerequisite for this exercise.

1. Sign in to the Azure Portal and navigate to your Azure Synapse Analytics workspace.

2. On your Azure Synapse workspace, click on Launch Synapse Studio, which will launch the Synapse Studio for you. From Synapse Studio, click on Data Hub from the left-hand navigation. You should have Data Object Explorer on your screen after this step.

3. Now, we will connect Azure Cosmos DB as a linked service. You can use Serverless Synapse SQL or Synapse Spark to explore, read, and write data into Azure Cosmos DB.

   a. Click on the + icon and then click on Connect External Data.

   b. Select the API to which you want to connect—SQL API or Mongo DB API—and click the Continue button.

   c. Use an appropriate name for the linked service and then select the Azure Cosmos DB account name and the database name.

   d. Click Create, and it will create a linked service that will allow you to connect to Azure Cosmos DB.

4. Connect Azure Cosmos DB database will show up in Linked tab under Azure Cosmos DB section.

5. You can easily select either OLTP-only Container or HTAP-enabled Container from the screen.

6. You can right-click on any container and you will be able to see a list of actions. Upon selection of any of those actions, it will have pre-configured code snippets that can be executed using either Synapse Spark Pool or Serverless Synapse SQL Pool.

Summary

Azure Synapse Link is the most exciting and advanced feature of Azure Synapse Analytics. It is a cloud-native hybrid transactional and analytical processing (HTAP) option available with Azure Cosmos DB. It is necessary to understand the OLTP and OLAP concepts, which are traditionally used for data analytics scenarios with any of the data lake, data warehouse, or data lakehouse implementations. We looked at both concepts in some detail and then jumped to explaining HTAP, which is the core concept for Azure Synapse Link. HTAP provides a tight and seamless integration between Azure Cosmos DB and Azure Synapse Analytics.

There are many benefits of using HTAP. The most important benefit is its no-ETL analytics architecture. HTAP does not require you to design, build, deploy, and maintain the data ingestion pipeline using an ETL tool. It simply and automatically syncs the data between operational store and analytical store. It saves you a lot of time and energy that go into building those complex data ingestion pipelines. It also helps in getting instant insights from your operational data. It reduces the technical and architectural complexity of your data analytics design. As there is no ETL required, it also helps in reducing data duplication, if not avoids it completely in many scenarios.

HTAP implementation in Azure Synapse Analytics is known as Azure Synapse Link. It supports Azure Cosmos DB as of now, which is in preview at the time of writing this book. Going forward, expect some more databases to get support for Azure Synapse Link. We looked at Azure Cosmos DB at a high level. Azure Cosmos DB is a multi-model truly globally distributed database. We also discussed two different data stores from Azure Cosmos DB, namely Azure Cosmos DB transactional store and Azure Cosmos DB analytical store. Later, we looked at the analytical store in a little more detail as it is a core part of Azure Synapse Link.

The Azure Cosmos DB analytical store provides many features that are good to know for understanding Azure Synapse Link. Hence, we discussed some of the important features of the Azure Cosmos DB analytical store, including columnar storage, decoupling of operational store, automatic data synchronization between Azure Cosmos DB transactional store and Azure Cosmos DB analytical store, and SQL API and Mongo DB API, which are currently supported for Azure Synapse Link.

Toward the end of the chapter, we discussed some of the real-life industry-specific use cases in which Azure Synapse Link can play an important role. First, we discussed an industrial IoT use case in detail, in which we discussed predictive maintenance pipelines and operational reporting. Azure Synapse Link can really help to predict the maintenance schedule for crucial industrial machinery very accurately, as we can use an anomaly detector machine learning algorithm trained with the latest data to make the prediction as accurate as possible so as to avoid unscheduled breakdowns of crucial industrial machinery laced with numerous industrial IoT device sensors. Operational reporting achieved by using Power BI–based dashboards allows you to generate various historical trends from the vast amount of sensor data.

The second use case we discussed involves real-time personalization for e-commerce users. Personalization based on the individual customer's preferences, order history, and so forth plays a crucial role in delivering customized products and services to users and businesses. The Recommendation API is used to recommend the most appropriate and personalized products and services to customers and internally uses a machine learning model trained using Synapse Spark on data stored in the Azure Cosmos DB analytical store.

With the end of this chapter, we are nearing the end of the book, but before we close, we have one more chapter in which we will discuss some of the use cases for Azure Synapse Analytics. We are also going to discuss some reference architecture that you can use to implement Azure Synapse Analytics. So, see you in the next and last chapter of this book!

CHAPTER 10

Azure Synapse Analytics Use Cases and Reference Architecture

This is the last chapter of this book. We have covered many topics about Azure Synapse Analytics in the previous chapters, which will help you begin your journey toward using Azure Synapse Analytics. As mentioned in other chapters, Azure Synapse Analytics is an amalgamation of multiple tools and technologies, so it is a little difficult to understand its architecture and its core components. Therefore, we have picked up each of those core components and discussed them in detail in individual chapters. We have covered Synapse SQL, Synapse Spark, Synapse Pipelines, Synapse Link, Synapse Workspace, and Synapse Studio in a good level of detail.

If you have gone through each of the previous nine chapters, then you should now have a very detailed idea of Azure Synapse Analytics as a comprehensive and powerful data analytics platform that combines the best of both worlds—Big Data analytics and the enterprise data warehouse. Therefore, in this chapter, we are going to discuss certain scenarios in which you should use Azure Synapse Analytics and certain other scenarios in which you should *not* use it. It is also necessary to understand the important use cases for Azure Synapse Analytics across various industries, like financial services, manufacturing, retail, and healthcare.

Toward the end of the chapter, we are going to discuss a couple of reference architectures in detail. Since Azure Synapse Analytics brings Big Data analytics and enterprise data warehouses together, we are going to discuss reference architecture for the modern data warehouse as well as real-time analytics on Big Data architecture in detail. We will look at architecture diagrams followed by data flows and the components used in both. So, let us get started.

225

© Bhadresh Shiyal 2021
B. Shiyal, *Beginning Azure Synapse Analytics*, https://doi.org/10.1007/978-1-4842-7061-5_10

Where Should You Use Azure Synapse Analytics?

Azure Synapse Analytics is a complex platform to understand as it contains multiple tools and technologies within it. Due to its architectural complexity and relative newness in the market, it is necessary to know where you should use Azure Synapse Analytics so that you get the desired benefits and outcomes for your projects. There are no hard and fast rules around it, but we can certainly follow certain general rules or best practices. So, let us discuss some scenarios in which it is best to use Azure Synapse Analytics.

Large Volume of Data

When you must store a large volume of data, you should think of using Azure Synapse Analytics. It can easily handle petabyte-scale data as it is built with this in mind. Azure Synapse Analytics supports Azure Data Lake Storage Gen 2 and Azure Blob Storage as the data storage layer, along with a traditional data warehouse that supports storing a large volume of structured data. Big Data analytics (BDA) workloads are big in data size, and a traditional tool or platform may not be able to handle the huge volume of data, while Azure Synapse Analytics can do that for you easily.

Disparate Sources of Data

Data integration will be a serious challenge to tackle when you have to deal with disparate data sources in order to acquire data for your data analytics workload. You will need to not only integrate and ingest the data from disparate data sources, but also store them at the single central location. In this scenario, Azure Synapse Analytics can easily meet your requirements. You can connect to disparate data sources and ingest the data into a central data lake storage from which you can clean and transform the data before loading it into the data warehouse. Azure Synapse Analytics supports these scenarios out of the box.

Data Transformation

If you need to clean, shape, transform, and consolidate data after ingesting it at a large scale then you can leverage Azure Synapse Analytics. Data cleansing and transformations are very obvious requirements for many of the data analytics workloads.

As you ingest data from disparate data sources in various data formats, it is obvious that you will need the capability to clean and transform those data before you can generate any insights from them. Azure Synapse Analytics supports Synapse SQL Pool as well as Synapse Spark Pool, which are scalable compute engines that can easily support these scenarios.

Batch or Streaming Data

Be it batch data ingestion or streaming data ingestion, both are supported in Azure Synapse Analytics. Batch data ingestion will integrate disparate data sources, from which data ingestion will be carried out in scheduled batches. However, if you have streaming data coming from clickstreams, IoT devices, sensors, and so forth, those can also be easily ingested into Azure Synapse Analytics. Synapse Pipelines within Azure Synapse Analytics supports batch data ingestion, while for streaming data ingestion, you can use Azure HDInsight with Apache Kafka. We will discuss a reference architecture for real-time analytics on Big Data architecture in detail later in this chapter.

Where Should You *Not* Use Azure Synapse Analytics?

In the previous section, we discussed where you should use Azure Synapse Analytics. We discussed some of the common scenarios in which it is recommended to use it. However, it is equally important to know in which scenarios Azure Synapse Analytics should not be used. Let us discuss those scenarios briefly.

Operational workloads, which are typically known as Online Transaction Processing (OLTP) systems, are not suitable to run on Azure Synapse Analytics. As it is a data analytics platform, it should be obvious that it is not a good idea to run any transaction-oriented systems on Azure Synapse Analytics, which is meant for Online Analytical Processing (OLAP) workloads.

Operational workloads tend to be very read and write intensive by nature and so are not an ideal match for Azure Synapse Analytics. Therefore, any workloads that include high-frequency reads and writes should not be considered for implementation using Azure Synapse Analytics.

Generally, bulk inserts through batch data ingestion or inserts through steaming data are supported in Azure Synapse Analytics, but if you have large volumes of single-row inserts happening very frequently then it is not a good scenario to cover in Azure Synapse Analytics. Similarly, if you are going to run singleton select queries, do not use Azure Synapse Analytics.

Azure Synapse Analytics is a combination of multiple tools and technologies, so it is necessary that you evaluate each of your scenarios carefully to decide which components of Azure Synapse Analytics you want to use in your project. It is not necessary to use each component from Azure Synapse Analytics, as you have separate pricing available for the majority of the main components.

For example, you may decide to use Synapse Pipelines and Azure Data Lake Storage from Azure Synapse Analytics, but you may not have a use case to justify the use of either Synapse SQL Pool or Synapse Spark Pool for your architecture. Similarly, you may prefer to use one of the other ETL/ELT tools to build your data pipelines and bring data inside Azure Synapse Analytics rather than using Synapse Pipelines, which is absolutely fine. The rule of thumb is that if you have a business scenario that can be aided by utilizing Azure Synapse Analytics' full potential then it is fine to utilize all or many of its components, but if you need just ad hoc querying capabilities from Serverless Synapse SQL Pools, as an example, then you do not need to utilize all the other components from Azure Synapse Analytics in your project.

Use Cases for Azure Synapse Analytics

Azure Synapse Analytics can be leveraged fully or partially for many industry-specific use cases. We will discuss these at a high level to clearly understand how Azure Synapse Analytics can help to solve any industry-specific use case.

Financial Services

Generally, financial services organizations deal with very large sets of data, including structured, unstructured, and semi-structured data, from various disparate data sources. Azure Synapse Analytics can provide a modern approach to managing Big Data within the financial services organization. It can also provide the opportunity to curate personalized customer experiences with effective compliance and governance practices to protect the customer data, which is of paramount importance for financial services. To deliver these personalized experiences, you can connect disparate data sources to create 360-degree customer profiles and curate exceptional experiences across every touchpoint, as financial services are delivered to the customer through multiple channels.

Fraud detection and risk management are other use cases for which you can utilize Azure Synapse Analytics. You can build customer trust with an end-to-end analytics solution that lets you monitor activities 24 hours a day across accounts, devices, and channels. This helps in detecting fraud at the first instance. For risk management, you can mitigate threats with a consolidated and flexible approach to collecting and analyzing enterprise data. You can develop cost-effective solutions using Azure Synapse Analytics that enable a rapid and actionable response to risks, which is a core function of risk management in financial services.

Manufacturing

Industry 4.0 combines operational and analytical technologies and galvanizes real-time access to new and existing data. Azure Synapse Analytics can help you to get real-time insights into your data for the manufacturing industry. Azure Synapse Analytics can help you to maximize overall equipment effectiveness. It can help you to improve factory productivity by maximizing availability, performance, and scalability. Azure Synapse Analytics can help you to generate powerful insights using advanced analytics with limitless scale.

By using the machine learning capabilities of Azure Synapse Analytics, you can predict asset failures so as to avoid costly downtime, reduce maintenance costs, and improve operational efficiency. With Synapse Link, you can apply advanced analytics and machine learning models with no data movement to optimize predictive maintenance. For the manufacturing industry, the supply chain plays a crucial role. By using Azure Synapse Analytics, you can connect disparate data sources throughout the supply chain to get a complete view of your business. By implementing advanced analytics and machine learning using Azure Synapse Analytics, you can improve visibility to increase resilience and gain a competitive advantage.

Retail

For the retail industry, customer service and the supply chain play crucial roles. You can leverage Azure Synapse Analytics to unify data from multiple channels and try to discover real-time insights with an end-to-end analytics service that helps you to know your customers and create a resilient supply chain. For the retail industry, personalized recommendations are very important to drive sales. You can use Azure Synapse Analytics to ingest, process, and analyze data to generate critical insights that help to provide better customer service. Using Azure Synapse Analytics, you can consolidate disparate customer data sources and process the data in real time to get a complete view of your customer.

The retail industry provides multiple platforms, channels, and devices via which the customers consume the retail services. By using Azure Synapse Analytics, you can provide a seamless customer journey with omnichannel optimization. This enables the customer to switch platforms, channels, and devices while still maintaining that same cohesive customer experience. You can use Azure Synapse Analytics to detect trends from customer browsing and post-purchase behavior to inform merchandizing decisions. You can also leverage Azure Synapse Analytics to eliminate any data silos with end-to-end analytics.

Healthcare

Due to the recent pandemic, the healthcare industry has gotten well-deserved attention. The industry faces pressure from changing patient expectations, regulatory constraints, and a shortage of care workers. You can leverage Azure Synapse Analytics to deliver personalized care, protect health-related data, and empower care teams to make the right decisions at the right time. Azure Synapse Analytics can be used to personalize care by providing patients with access to the health data they need to get the right care and attention at the right time. It can help you to optimize operations by aggregating data across various health IT systems and automating routine operations.

You can also implement clinical analytics by using Azure Synapse Analytics. You can build a system that can analyze patient data to associate it with symptoms of various diseases and recommend treatment protocols. In the healthcare industry compliance is also an important requirement. You can apply compliance analytics by identifying regulatory risks and workflow efficiencies across datasets. Azure Synapse Analytics can also help you to keep sensitive health data secure and private at all times.

Reference Architectures for Azure Synapse Analytics

A reference architecture provides recommended integrations of IT tools and services and proven structures to articulate a solution. Generally, a reference architecture symbolizes best practices that are accepted in the industry and suggests the optimal delivery mechanism for a set of specific technologies. In the case of Azure, you will have Azure reference architectures, which will provide you with insight into how you can build your own architecture; they are well documented and available from Microsoft.

A reference architecture helps all stakeholders—including project managers, development team, solution architect, and so on—to understand and collaborate effectively to implement a specific project. Basically, it is an architecture you can refer to and understand easily to see if you can directly implement the full reference architecture or perhaps just a specific service or component from it. Microsoft provides a good level of documentation around Azure architecture, which is available at the Azure Architecture Center. This provides robust guidance for designing solutions on Azure by using established best patterns and practices. You can visit it at `https://docs.microsoft.com/en-us/azure/architecture/`.

Each reference architecture in the Azure Architecture Center includes some recommended practices along with considerations for resiliency, availability, security, scalability, and other aspects of the design. There are many reference architectures available that involve Azure Synapse Analytics as one of the components. Here, we are going to discuss a couple of reference architectures that are related to Azure Synapse Analytics.

Modern Data Warehouse Architecture

A modern data warehouse lets you easily bring together all your data at any scale, and to get insights through analytical dashboards, operational reports, or advanced analytics for all your users. We looked at modern data warehouses and data lakehouses in detail in Chapter 2. Modern data warehouse architecture includes multiple services like integration services, data processing services, data storage services, data analysis services, and data visualization services for an end-to-end modern data warehouse implementation. Azure Synapse Analytics has many of these services built into it, which eases the usability and operational aspects of implementation. This also simplifies the overall architecture of the solution to be implemented with this reference architecture.

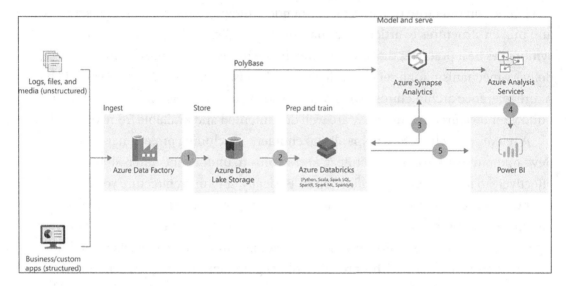

Figure 10-1. *Modern data warehouse reference architecture. Source: Microsoft. This image is taken from Microsoft's Azure Architecture Center for the reference architecture titled "Modern Data Warehouse Architecture."* `https://docs. microsoft.com/en-us/azure/architecture/solution-ideas/articles/ modern-data-warehouse`

Let us look at the data flow for the reference architecture depicted in Figure 10-1.

1. As part of the modern data warehouse architecture, we will have to ingest the data from the various data sources into Azure Data Lake Storage. Therefore, we will need a data integration service to ingest the data for us. Azure Data Factory can help us to meet the data ingestion requirements. It will help us to combine all our structured, unstructured, and semi-structured data, including various files, logs, audio, video, and so forth.

2. Once the data is ingested and stored in Azure Data Lake Storage, we can leverage the Azure Databricks service to clean and transform the data for data analytics purposes. We can also think of creating various zones or layers within Azure Data Lake Storage for storing the intermediate data as part of data analytics processing. For example, we can read the data from the raw zone, which stores the data in its original raw format, clean the data using Azure Databricks, and store it in the landing zone. Here also, for data pipeline orchestration, Azure Data Factory can be leveraged easily.

3. Once we have cleaned and transformed data, it can be moved to Azure Synapse Analytics to merge with the existing structured data. This will give us a central hub to store and retrieve a single version of the truth. Here, we can leverage the existing connector, which allows you to connect to Azure Databricks from Azure Synapse Analytics.

4. To build operational reports and analytical dashboards to derive deeper insights from your data, you can leverage Azure Analysis Service to serve a huge number of end users who are consuming those operational reports and analytical dashboards. Azure Analysis Service can connect with Azure Synapse Analytics and keep the frequently used data in memory to serve multiple end users with a decent response time. The end users can consume these reports via Power BI Service, which is a widely used data visualization tool.

5. Power BI also supports directly querying data from Azure Databricks through a dedicated connector. So, you can utilize that option to run the ad hoc queries directly against Azure Databricks. So, you can bypass Azure Synapse Analytics and Azure Analysis Service for ad hoc queries.

Now let us look at the various services and components that are part of this reference architecture. Along with that, we will also look at any alternative services or components we can use instead of the ones used in the reference architecture.

1. **Azure Data Factory:**

   Azure Data Factory is a hybrid data integration and orchestration service that allows you to design, develop, deploy, schedule, orchestrate, and monitor various data pipelines or ETL/ELT workflows very easily. It has multiple different connectors that allow you to connect to various sources and sinks with very quick turnaround time.

   You also have the option to use Synapse Pipelines here instead of Azure Data Factory. Synapse Pipelines are almost similar to Azure Data Factory, and it has all the features you would get with Azure Data Factory. As Synapse Pipelines are directly integrated in Azure Synapse Analytics, you can design, develop, deploy, schedule, orchestrate, and monitor the data pipelines directly from Synapse Studio without leaving Azure Synapse Analytics. However, if you use Azure Data Factory, then it will be same interface, but it will not be directly accessible from Azure Synapse Analytics.

2. **Azure Data Lake Storage:**

   Azure Data Lake Storage (ADLS) is massively scalable storage. It supports hierarchical namespaces, which enables you to define your permission levels granularly. ADLS allows you to store various structured, semi-structured, and unstructured data formats easily. That means that you can store all kinds of files, like images, videos, documents, audios, and so forth, in Azure Data Lake Storage.

   Here, you have the option to use the Azure Data Lake Storage account that is already linked with your Azure Synapse Analytics account. We know that Azure Synapse Analytics supports Azure Data Lake Storage as one of the storage options. However, an ADLS account not linked to your Azure Synapse Analytics workspace can also be used.

In the end, though, it makes more sense to leverage the already
linked Azure Data Lake Storage account for modern data warehouse
architecture. It will provide a more seamless experience due to its
tighter integration with Azure Synapse Analytics.

3. **Azure Databricks:**

Azure Databricks is an easy-to-use, fast, and highly collaborative
data analytics platform. It is also based on Apache Spark.
Databricks is an independent company founded and managed
by the inventor of Apache Spark. Hence, the implementation of
Apache Spark in Databricks does perform better and faster than a
normal open source version of Apache Spark. Azure Databricks is
generally used to run jobs and notebooks containing medium to
complex data transformations at scale on large amounts of data.

Alternatively, you have the option to use Synapse Spark from Azure
Synapse Analytics instead of Azure Databricks. We discussed Synapse
Spark in detail in Chapter 6. It is Microsoft's own implementation of
Apache Spark and is in direct competition to Databricks. It provides
many features similar to what you get in Azure Databricks.

4. **Azure Analysis Service:**

Azure Analysis Service is a fully managed Platform as a Service
(PaaS) that provides enterprise-grade data models in the cloud.
It allows you to use advanced mesh-up and modeling features
to combine data from multiple data sources, define metrics, and
secure your data in a single trusted, tabular, semantic data model.
Data models deployed to Azure Analysis Service provide an easier
and faster way for users to carry out data analysis work using tools
like Power BI and even an Excel sheet.

There seems to be no direct alternative within Azure that you can
use instead of Azure Analysis Service, but Power BI does have
some data modeling capabilities that can be explored to know if
that meets all the requirements around data modeling or not. If
it meets the requirements, then you can use Power BI instead of
Azure Analysis Service for data modeling purposes.

5. **Power BI:**

   Power BI is a suite of business analytics tools that delivers
   insights throughout your organization. It allows you to connect to
   hundreds of data sources, simplify data preparation, and drive ad
   hoc analysis. Power BI produces beautiful reports, then publishes
   them for your organization to consume on the web and across
   mobile devices.

   There is no direct alternative for Power BI within the Azure
   suite of services, but you can link your Power BI workspace to
   Azure Synapse Analytics from Synapse Studio. That way you get
   seamless integration with Azure Synapse Analytics, and it will
   increase collaboration among the different teams and will result in
   quicker go-to-market time.

To summarize, it is recommended that you use the components available within
Azure Synapse Analytics for better collaboration and a seamless development
experience compared to using independent Azure services like Azure Data Lake Storage,
Azure Data Factory, and Azure Databricks. Hence, this reference architecture can further
be enhanced by accommodating all these Azure Synapse Analytics options.

Now, let us move on to our second reference architecture for Azure Synapse
Analytics.

Real-Time Analytics on Big Data Architecture

Azure Synapse Analytics brings the best of both worlds—Big Data analytics (BDA)
and enterprise data warehouse (EDW)—together. In the previous section, we saw how
Azure Synapse Analytics can be leveraged effectively for implementing enterprise
data warehouse workloads. Now, in this section, we are going to look at how Big Data
analytics–related workloads can be handled by Azure Synapse Analytics.

Real-time analytics always poses challenges as it has to deal with a continuous
stream of data coming from various disparate data sources, and the proposed
architecture will have to take care of processing it in real time. It also means that you will
end up handling a huge amount of data in your system. So, this reference architecture
has been designed to handle real-time analytics on Big Data. Let us look at the data flow
for this reference architecture, which is shown in Figure 10-2.

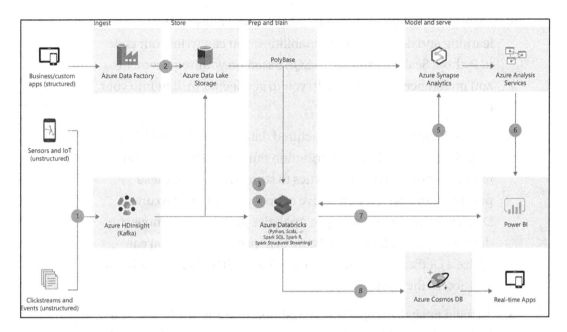

Figure 10-2. *Real-time analytics on Big Data reference architecture. Source: Microsoft. This image is taken from Microsoft's Azure Architecture Center for the reference architecture titled "Real Time Analytics on Big Data Architecture."* `https://docs.microsoft.com/en-us/azure/architecture/solution-ideas/ articles/real-time-analytics`

1. Unstructured data from sensors and IoT devices as well as clickstreams and events are to be ingested using live streaming. An Apache Kafka cluster hosted on Azure HDInsight will bring this capability to this reference architecture to achieve this goal.

2. Structured data from business applications and custom applications can easily be ingested using Azure Data Factory into Azure Data Lake Storage. Here, data coming via livestream from Apache Kafka, running on Azure HDInsight, will be combined with structured data ingested using Azure Data Factory.

3. To combine the live-streaming unstructured data with structured data, you will have to analyze, clean, and transform the data. You can leverage Azure Databricks to provide data processing capabilities; it can be used for both batch processing and stream processing.

237

4. You can also leverage Azure Databricks' scalable machine learning and deep learning capabilities. You can write your code in Python, Scala, R, and Spark SQL based on your requirements and preferences. This will help you to get deeper insight into your data.

5. Cleaned and transformed structured data need to be stored for modeling and serving or distribution purposes. Here, you can leverage Azure Synapse Analytics to store the data for these purposes. You can rely on native connectors between Azure Databricks and Azure Synapse Analytics to access and move the data at a large scale. Within Azure Synapse Analytics, you can utilize a Dedicated or Provisioned Synapse SQL Pool to load, store, and access the data.

6. To build analytical reports and dashboards from the data stored in Azure Synapse Analytics, you will need to model the data and make aggregated and consolidated views available to end users via data visualization tools. Here, for modeling, you can use Azure Analysis Services, and for data visualization, Power BI is a good choice.

7. Advanced users of Power BI will be able to take advantage of a direct connector between Power BI, Azure Databricks, and Azure HDInsight to perform root-cause determination and raw data analysis.

8. For real-time applications, you can leverage Azure Cosmos DB to connect to Azure Databricks to generate real-time insights from your data.

Now, let us discuss the components or services that are used in the preceding reference architecture for real-time analytics on Big Data. Some of the components were seen in the previous reference architecture; hence, we are not going to discuss those components here. You can refer to the previous reference architecture if required. So, we are left with only two new components or services—Azure HDInsight and Azure Cosmos DB.

1. **Azure HDInsight:**

   Azure HDInsight is a fully managed, full-spectrum open source analytics service for popular open source frameworks. It includes Apache Hadoop, Apache Spark, Apache Hive, LLAP (Live Long and Process), Kafka, Storm, R, and others. Azure HDInsight can effortlessly process massive amounts of data and get all the benefits of the broad open source project ecosystem with the global scale provided by Azure services. It is a good option via which to easily migrate your on-premises workloads to the cloud.

2. **Azure Cosmos DB:**

   Azure Cosmos DB is a fully managed NoSQL database service. It gives you guaranteed single-digit millisecond response times and 99.999 percent availability, backed by SLAs, automatic and instant scalability, and open source APIs for MongoDB and Cassandra. You can get very fast writes and reads anywhere in the world with turnkey data replication and multi-region writes. It allows you to gain insight over real-time data with no-ETL analytics using Azure Synapse Link for Azure Cosmos DB. Here, we are using this in the reference architecture to provide real-time data to real-time applications. Internally, it relies on a connector from Azure Cosmos DB to Azure Databricks to provide real-time data updates to the consuming application.

Summary

Azure Synapse Analytics is a complex data analytics platform to understand and learn. It is also a relatively new platform. There are many use cases for Azure Synapse Analytics, and similarly many different reference architectures.

It is important to know where you should use Azure Synapse Analytics and where you should not use it. Azure Synapse Analytics is a good fit when you have a very large amount of data, disparate sources of data, large-scale data transformations, and batch or stream data ingestion as part of your business requirements. Azure Synapse Analytics is a combination package that supports both Big Data analytics and enterprise data warehouse–related workloads.

There are certain anti-patterns or scenarios in which you should not be using Azure Synapse Analytics. It makes sense to know these scenarios so that you can avoid mistakes while you implement Azure Synapse Analytics in your project. Azure Synapse Analytics is not a good fit for any Online Transaction Processing system (OLTP). If your system is going to be very write or read intensive, then it is not a good idea to leverage Azure Synapse Analytics in your project. We discussed a few of these scenarios in which you should not implement Azure Synapse Analytics. We also discussed that it is not necessary to utilize all the components or tools from Azure Synapse Analytics. Based on your business requirements, you should decide if you want to utilize Azure Synapse Analytics fully or partially.

We discussed some of the industry-specific use cases of Azure Synapse Analytics. We noted some use cases from financial services and looked at how Azure Synapse Analytics can be leveraged to deliver personalized experiences, perform fraud detection at first instance, and manage risk for financial services customers. The manufacturing industry also has many use cases in which Azure Synapse Analytics can be leveraged. For example, you can maximize overall equipment effectiveness, reduce unplanned downtime, and enhance supply-chain visibility by using Azure Synapse Analytics. The retail industry also has many use cases that can be implemented using Azure Synapse Analytics, which include personalized recommendations, use of omnichannel optimization, and building a resilient supply chain. We also discussed some of the use cases for the healthcare industry like personalized care, clinical analytics, healthcare compliance analytics and so forth.

Toward the end of the chapter, we discussed reference architectures for Azure Synapse Analytics. Reference architectures are well documented and publicly available on the Azure Architecture Center website. It is easy to refer to those architectures to validate your own architecture against it or to build a new architecture by referring to the reference architecture. As part of the reference architecture discussion, we examined two important reference architectures of Azure Synapse Analytics.

We discussed the modern data warehouse–related reference architecture in detail and looked at its data flow as well as various components, like Azure Data Factory, Azure Data Lake Storage, Azure Databricks, Azure Analysis Service, and Power BI. We also discussed some of the alternative components that can be leveraged instead of the one proposed in the reference architecture. Similar to that, we also discussed another reference architecture related to real-time analytics on Big Data architecture. In that reference architecture, we looked at the overall data flow and its components in detail.

With this chapter, this book is complete. I hope that after reading this book you are on a solid foundation to kickstart or begin your journey into the amazing world of Azure Synapse Analytics.

Thanks for reading!

Index

A

© Bhadresh Shiyal 2021
B. Shiyal, *Beginning Azure Synapse Analytics*, https://doi.org/10.1007/978-1-4842-7061-5

T

Printed in the United States
by Baker & Taylor Publisher Services